Orphan Black and Philosophy

Popular Culture and Philosophy® Series Editor: George A. Reisch

Volume 1 *Seinfeld and Philosophy: A Book about Everything and Nothing* (2000)

Volume 2 *The Simpsons and Philosophy: The D'oh! of Homer* (2001)

Volume 3 *The Matrix and Philosophy: Welcome to the Desert of the Real* (2002)

Volume 4 *Buffy the Vampire Slayer and Philosophy: Fear and Trembling in Sunnydale* (2003)

Volume 9 *Harry Potter and Philosophy: If Aristotle Ran Hogwarts* (2004)

Volume 12 *Star Wars and Philosophy: More Powerful than You Can Possibly Imagine* (2005)

Volume 13 *Superheroes and Philosophy: Truth, Justice, and the Socratic Way* (2005)

Volume 19 *Monty Python and Philosophy: Nudge Nudge, Think Think!* (2006)

Volume 25 *The Beatles and Philosophy: Nothing You Can Think that Can't Be Thunk* (2006)

Volume 30 *Pink Floyd and Philosophy: Careful with that Axiom, Eugene!* (2007)

Volume 35 *Star Trek and Philosophy: The Wrath of Kant* (2008)

Volume 36 *The Legend of Zelda and Philosophy: I Link Therefore I Am* (2008)

Volume 39 *Jimmy Buffett and Philosophy: The Porpoise Driven Life* (2009) Edited by Erin McKenna and Scott L. Pratt

Volume 42 *Supervillains and Philosophy: Sometimes Evil Is Its Own Reward* (2009)

Volume 45 *World of Warcraft and Philosophy: Wrath of the Philosopher King* (2009) Edited by Luke Cuddy and John Nordlinger

Volume 46 *Mr. Monk and Philosophy: The Curious Case of the Defective Detective* (2010) Edited by D.E. Wittkower

Volume 48 *The Red Sox and Philosophy: Green Monster Meditations* (2010) Edited by Michael Macomber

Volume 49 *Zombies, Vampires, and Philosophy: New Life for the Undead* (2010) Edited by Richard Greene and K. Silem Mohammad

Volume 51 *Soccer and Philosophy: Beautiful Thoughts on the Beautiful Game* (2010) Edited by Ted Richards

Volume 52 *Manga and Philosophy: Fullmetal Metaphysician* (2010) Edited by Josef Steiff and Adam Barkman

Volume 53 *Martial Arts and Philosophy: Beating and Nothingness* (2010) Edited by Graham Priest and Damon Young

Volume 54 *The Onion and Philosophy: Fake News Story True, Alleges Indignant Area Professor* (2010) Edited by Sharon M. Kaye

Volume 55 *Doctor Who and Philosophy: Bigger on the Inside* (2010) Edited by Courtland Lewis and Paula Smithka

Volume 57 *Rush and Philosophy: Heart and Mind United* (2011) Edited by Jim Berti and Durrell Bowman

Volume 58 *Dexter and Philosophy: Mind over Spatter* (2011) Edited by Richard Greene, George A. Reisch, and Rachel Robison-Greene

Volume 60 *SpongeBob SquarePants and Philosophy: Soaking Up Secrets Under the Sea!* (2011) Edited by Joseph J. Foy

Volume 61 *Sherlock Holmes and Philosophy: The Footprints of a Gigantic Mind* (2011) Edited by Josef Steiff

Volume 62 *Inception and Philosophy: Ideas to Die For* (2011) Edited by Thorsten Botz-Bornstein

Volume 63 *Philip K. Dick and Philosophy: Do Androids Have Kindred Spirits?* (2011) Edited by D.E. Wittkower

Volume 64 *The Rolling Stones and Philosophy: It's Just a Thought Away* (2012) Edited by Luke Dick and George A. Reisch

Volume 67 *Breaking Bad and Philosophy: Badder Living through Chemistry* (2012) Edited by David R. Koepsell and Robert Arp

Volume 68 *The Walking Dead and Philosophy: Zombie Apocalypse Now* (2012) Edited by Wayne Yuen

Volume 69 *Curb Your Enthusiasm and Philosophy: Awaken the Social Assassin Within* (2012) Edited by Mark Ralkowski

Volume 71 *The Catcher in the Rye and Philosophy: A Book for Bastards, Morons, and Madmen* (2012) Edited by Keith Dromm and Heather Salter

Volume 72 *Jeopardy! and Philosophy: What Is Knowledge in the Form of a Question?* (2012) Edited by Shaun P. Young

Volume 73 *The Wire and Philosophy: This America, Man* (2013) Edited by David Bzdak, Joanna Crosby, and Seth Vannatta

Volume 74 *Planet of the Apes and Philosophy: Great Apes Think Alike* (2013) Edited by John Huss

Volume 75 *Psych and Philosophy: Some Dark Juju-Magumbo* (2013) Edited by Robert Arp

Volume 79 *Frankenstein and Philosophy: The Shocking Truth* (2013) Edited by Nicolas Michaud

Volume 80 *Ender's Game and Philosophy: Genocide Is Child's Play* (2013) Edited by D.E. Wittkower and Lucinda Rush

Volume 81 *How I Met Your Mother and Philosophy: Being and Awesomeness* (2014) Edited by Lorenzo von Matterhorn

Volume 82 *Jurassic Park and Philosophy: The Truth Is Terrifying* (2014) Edited by Nicolas Michaud

Volume 83 *The Devil and Philosophy: The Nature of His Game* (2014) Edited by Robert Arp

Volume 84 *Leonard Cohn and Philosophy: Various Positions* (2014) Edited by Jason Holt

Volume 85 *Homeland and Philosophy: For Your Minds Only* (2014) Edited by Robert Arp

Volume 86 *Girls and Philosophy: This Book Isn't a Metaphor for Anything* (2014) Edited by Richard Greene and Rachel Robison-Greene

Volume 87 *Adventure Time and Philosophy: The Handbook for Heroes* (2015) Edited by Nicolas Michaud

Volume 88 *Justified and Philosophy: Shoot First, Think Later* (2015) Edited by Rod Carveth and Robert Arp

Volume 89 *Steve Jobs and Philosophy: For Those Who Think Different* (2015) Edited by Shawn E. Klein

Volume 90 *Dracula and Philosophy: Dying to Know* (2015) Edited by Nicolas Michaud and Janelle Pötzsch

Volume 91 *Its Always Sunny and Philosophy: The Gang Gets Analyzed* (2015) Edited by Roger Hunt and Robert Arp

Volume 92 *Orange Is the New Black and Philosophy: Last Exit from Litchfield* (2015) Edited by Richard Greene and Rachel Robison-Greene

Volume 93 *More Doctor Who and Philosophy: Regeneration Time* (2015) Edited by Courtland Lewis and Paula Smithka

Volume 94 *Divergent and Philosophy: The Factions of Life* (2016) Edited by Courtland Lewis

Volume 95 *Downton Abbey and Philosophy: Thinking in That Manor* (2016) Edited by Adam Barkman and Robert Arp

Volume 96 *Hannibal Lecter and Philosophy: The Heart of the Matter* (2016) Edited by Joseph Westfall

Volume 97 *The Ultimate Walking Dead and Philosophy: Hungry for More* (2016) Edited by Wayne Yuen

Volume 98 *The Princess Bride and Philosophy: Inconceivable!* (2016) Edited by Richard Greene and Rachel Robison-Greene

Volume 99 *Louis C.K. and Philosophy: You Don't Get to Be Bored* (2016) Edited by Mark Ralkowski

Volume 100 *Batman, Superman, and Philosophy: Badass or Boyscout?* (2016) Edited by Nicolas Michaud

Volume 101 *Discworld and Philosophy: Reality Is Not What It Seems* (2016) Edited by Nicolas Michaud

Volume 102 *Orphan Black and Philosophy: Grand Theft DNA* (2016) Edited by Richard Greene and Rachel Robison-Greene

Volume 103 *David Bowie and Philosophy: Rebel Rebel* (2016) Edited by Theodore G. Ammon

In Preparation:

Deadpool and Philosophy (2016) Edited by Nicolas Michaud

Peanuts and Philosophy (2016) Edited by Richard Greene and Rachel Robison-Greene

Red Rising and Philosophy (2016) Edited by Courtland Lewis and Kevin McCain

The X-Files and Philosophy (2016) Edited by Robert Arp.

The Ultimate Game of Thrones and Philosophy (2016) Edited by Eric J. Silverman and Robert Arp

Hamilton and Philosophy (2017) Edited by Aaron Rabinowitz and Robert Arp.

X-Files and Philosophy (2017) Edited by Robert Arp.

Mr. Robot and Philosophy (2017) Edited by Richard Greene and Rachel Robison-Greene

For full details of all Popular Culture and Philosophy® books, visit www.opencourtbooks.com.

Popular Culture and Philosophy®

Orphan Black and Philosophy

Grand Theft DNA

EDITED BY

RICHARD GREENE AND
RACHEL ROBISON-GREENE

OPEN COURT

Volume 102 in the series, Popular Culture and Philosophy®, edited by George A. Reisch

To find out more about Open Court books, call toll-free 1-800-815-2280, or visit our website at www.opencourtbooks.com.

Open Court Publishing Company is a division of Carus Publishing Company, dba Cricket Media.

Copyright © 2016 by Carus Publishing Company, dba Cricket Media

First printing 2016

All rights reserved. No part of this publication may be reproduced, stored in a retrieval system, or transmitted, in any form or by any means, electronic, mechanical, photocopying, recording, or otherwise, without the prior written permission of the publisher, Open Court Publishing Company, 70 East Lake Street, Suite 800, Chicago, Illinois 60601.

Printed and bound in the United States of America.

Orphan Black and Philosophy: Grand Theft DNA

ISBN: 978-0-8126-9920-3

This book is also available as an e-book.

Library of Congress Control Number: 2016942074

For Jason Robison

Contents

Thanks

Working on this project has been a pleasure, in no small part because of the many fine folks who have assisted us along the way. In particular, a debt of gratitude is owed to David Ramsay Steele and George Reisch at Open Court, Graeme Manson and John Fawcett (of course!), the contributors to this volume, and our respective academic departments at UMass Amherst and Weber State University. Finally, we'd like to thank those family members, students, friends, and colleagues with whom we've had fruitful and rewarding conversations on various aspects of all things *Orphan Black* as it relates to philosophical themes.

Leda Thinkin' to Us . . .

Suppose that you find yourself in a subway station staring face to face with your own face. You're looking at your face, but it's not your face—your face is not on you. Unbeknownst to you, it is the face of one of your many clones.

Further imagine that your doppleganger commits suicide by hurling herself in front of a train, leaving her identity behind for you to take, if you dare. Most folks would have no reason to assume the identity of another person, but in this particular instance, you are also to suppose that your life is in a shambles, and you're on the run from your potentially homocidal, drug-dealing ex, whom you have just robbed. Under the circumstances, becoming someone else entirely seems kinda like a no-brainer.

This is exactly the situation that Sarah Manning found herself in, and exactly what she did. It took some fancy footwork, conniving, and a whole lot of pluck, but Sarah (at least for a while) pulled it off—she fooled her "new" co-workers, and her "new" boyfriend, and she escaped the wrath of her crazy ex-boyfriend. Unfortunately, it didn't work out all that well for Sarah. Over the following several months she was shot at (on may occasions!), imprisoned, extorted, kidnapped, tortured, lost her daughter, learned horrible things about her past, was involved in numerous altercations, and was generally a huge magnet for drama of every variety. Perhaps adopting the identity of another person was not the smartest move after all.

What should Sarah have done? It probably didn't seem to her as though she had many options, and she certainly could not have foreseen the impending shit-storm that was about to hit her. Still, there are always options.

What she probably should have done is read through this book, cover to cover! Then she would have known more about the metaphysics of cloning. She would have delved into the murky philosophical details of personal identity. She would have learned about the value of sisterhood, and the philosophical significance of family. She could have spent countless delightful hours contemplating the absurdity of it all, Buddhism, feminism, epistemology, and a host of other snazzy philosophical topics. Truly Sarah missed a great opportunity to seek what the philosopher John Stuart Mill called "higher pleasures." Fortunately, what happened to Sarah has yet to happen to you. You still have time to read the book before you find out that you're a scientific experiment of the highest order, and that your life is in great peril!

Quick, there's not a moment to lose. As your monitors, we'll guide you through it all, and of course, we'll be watching you closely. On to Chapter 1 . . .

PART I

"How many of us are there?"

1
Fearfully and Wonderfully Made

JOHN V. KARAVITIS

It starts like this. You've just lifted a bag of cocaine from your boyfriend, with the intention of having your foster brother flip it for some quick cash. An accomplished and street-savvy hustler, it's business as usual for you, as it has been since your youth, growing up an orphan in foster care.

Unwanted and angry at the world, you quickly learned that all that matters is to always look out for Number One. Your life is worth living, and you're strong enough to do whatever it takes to survive. But this time, one cool fall evening,[1] while walking along a train platform, you see *you* (!) turn around and look right back at you. That is, looking right at you just before stepping off of the train platform and right into the path of an oncoming train.

It doesn't take long for you to collect yourself. You are, of course, a creature of the streets. You're self-propelled, and tough as nails. And, as bizarre a situation as this is, you know enough to instinctively go for that woman's purse. The woman who just a few moments before had willingly stepped

---

[1] The events in the first three seasons of *Orphan Black* appear to take place across a span of about thirty to forty days. In "Formalized, Complex, and Costly," we see *Castor* clone Mark Rollins try to escape into what appears to be a mature corn field. In Canada, corn matures in October, and is harvested in November.

in front of an oncoming train. The woman whose identity you've decided to borrow. The now dead woman who looked *just like you*.

Sometimes, it's the simplest things, the thoughtless and habitual actions that we take as we navigate our way in the world, which can end up tripping us down into a rabbit hole. And some rabbit holes, we find out too late, are simply too deep for us to climb out of by ourselves.

Orphan Black presents the story of Sarah Manning, a clone whose life is turned upside-down as she learns the truth of her origin. We can't say for sure how her life would have turned out had she simply walked away from that train platform that evening. There's the chance that she would have lived out her existence without once coming into contact with any of her sister clones, or crossing paths with the Neolution cabal that now threatens her and her daughter Kira.

The existence of Sarah and the *Leda* and *Castor* clones raises many questions about procreative rights and responsibilities. More to the point, knowing that the *Leda* clones were intentionally and knowingly created to lack the ability to bear children, and also that they will all inevitably suffer from a deadly respiratory illness that is linked to their genetically planned infertility, raises the philosophical issue known as the "non-identity problem." This issue has also been referred to as the "paradox of future individuals." Since it was first raised in 1976 by contemporary philosopher Derek Parfit, it's been applied to environmental ethics, and has also been used to support what's known as antinatalism—the position that the human race should not bring forth any more children, and should die out.

So please, take my hand, gentle reader, and hold on tight as I guide us down into this philosophical rabbit hole. Don't let go—I promise you *we'll* find our way back out. Ready?

Conditions of Existence

Very few, if any, acts can be seen as more fraught with responsibility than the creation of another human being.

Human babies are born helpless, utterly dependent on their parents or some caregiver. It takes a long time to mature into an independent adult who can stand on their own two feet. To knowingly bring a child into a world where it would be in grave danger, or to bring forth a child who would in some way be at a disadvantage from the moment of its birth, would seem to violate the intuition of most people. It just seems *wrong* to make a decision to procreate in a situation where the best interests of the future child would not be ensured and protected. Wouldn't making such a decision cause harm to the future child?

This is the path that Sarah seems to have chosen when she bore Kira. Sarah became pregnant with Kira during her relationship with Cal Morrison. Twenty-one years old when Kira was born, Sarah still lacked an education, a steady means of employment, and a life partner who could have easily helped her shoulder the financial and emotional costs of raising a child. When Paul Dierden visits Cal Morrison at his new apartment, we learn that Cal is in fact very wealthy due to the work that he did designing weapon systems for the military ("Transitory Sacrifices of Crisis"). Sarah never told Cal that she was pregnant, but she also didn't elect to terminate her pregnancy.

Suppose the fact that Sarah bore Kira when she was only twenty-one and unable to properly raise her, which led her to leaving Kira with Siobhan Sadler, her own foster mother, causes Kira to eventually have problems relating to and trusting people. This situation in turn affects whether Kira can successfully participate in intimate relationships and eventually bear and raise children of her own. Sarah's actions could be seen as having been selfish, easily avoidable, and eventually harmful to her child. Had Sarah waited until she had been in a stronger financial and emotional position to raise a child, that child would probably not have suffered from those same relationship problems upon reaching maturity.

If Kira were to confront Sarah later in life and complain about the reduced quality of life that she suffers because of the circumstances of her creation, Sarah might very well reply to Kira that, had she waited, Kira would not be alive to complain about her reduced quality of life! In fact, Sarah could point out that, although life is difficult for Kira as an adult, it's not so bad that Kira lacks a life worth living. Even if Sarah's decision to have Kira when she did negatively affected the later quality of her life, nevertheless, had the exact sequence of steps that were taken not been taken, Kira would not be alive to complain about it. Kira's only alternative to her reduced quality of life, albeit a life that is still, overall, worth living, would be non-existence!

Kira may have indeed in some way been wronged by Sarah's selfish actions; and, intuitively, most people would agree. *But Sarah's selfish actions are the condition of Kira's existence.* Intuitively, Sarah does appear to have done Kira some wrong by bringing her into the world under such circumstances. By not having waited to have children until a time when she would have been better able to shoulder the financial and emotional costs, Sarah's actions do seem to have harmed Kira. Yet without those very same actions, in the exact sequence they were made, Kira would never have been born.

It was just such a situation that philosopher Derek Parfit raised in his 1984 book *Reasons and Persons*. Parfit asks us to consider a fourteen-year-old girl who decides to become pregnant. Assuming that there are no medical issues, this desire is viable. She can certainly go ahead and conceive and give birth. However, the child that she bears will be born to a teenage mother. Teenagers are not expected to have either the financial or emotional resources to adequately provide for a child. In a sense, bearing a child at her young age would disadvantage that child, setting it up for developmental, medical, and even social problems in its future. These future problems may not be either preventable or curable, even if foreseeable. This child's quality of life would be much lower than it could have been, even if, despite all of its future problems, the child would, overall, still be able to have a life worth living.

Wouldn't it have been better if this fourteen-year-old girl had waited to bear a child at a later point in life? Had this fourteen-year-old girl waited until she had graduated from college, become established in her career path, saved some money, and found a life partner with whom she could share the rest of her life, wouldn't this alternate timeline in fact have been better for both her and her future child?

It is obvious that a child born at one point in time is "not identical" to a child born later. Of course the two children would be different people! The child that Derek Parfit's hypothetical fourteen-year-old girl *actually had* would not be the same child that she *could have had* at a much later time in her life. It's the same with Sarah and Kira. At first glance, this seems nonsensical, that it's a dilemma created solely for the sake of argument.

No one would disagree that, regardless of how her life will turn out, Kira would more than likely still have had a life worth living. Yet, since she could not have been born any other way and still be the same Kira, *Sarah could not have caused Kira any harm*. This apparent contradiction between what we tend to suppose at first blush—that Sarah harmed Kira by deciding to conceive, bear, and raise her in less than advantageous circumstances—and the logical conclusion that Sarah has *not* harmed Kira because Kira could not have come into existence any other way, is known as the "Implausible Conclusion." It is the heart of Parfit's non-identity problem.

Ipsa Scientia Potesta Est

You may not be convinced, so I'll rephrase the problem, more in line with what we've seen in *Orphan Black*. There are many variations of this problem, but, suffice it to say, there are two scenarios presented. In the first, the parent is seen as acting in such a way that it is obvious that she has, either intentionally or through negligence, harmed her future child. In the second, although the parent takes action that results in a child that "comes into existence" in the exact same manner, and with the exact same degree of disadvantage or disability, it's not as clear that one can say with certainty that this second child has been harmed.

Suppose that Alice wants to conceive a child. Her obstetrician gives her the go-ahead, but warns her that, if she continues to take a specific recreational drug, her future child will be born deaf. For whatever reason, Alice continues to take that recreational drug during her pregnancy, and, as predicted, her child is born deaf. Suppose another woman, Carol, decides to use PGD (preimplantation genetic diagnosis) in order to select an embryo such that the resulting child will be deaf from birth. Feel free to consider Carol either deaf herself, and thus wanting a child that will also experience life as a member of the deaf community, or perhaps Carol feels that, by being born deaf, her future child will be better able to handle life's challenges.

The embryo that Carol chooses is implanted, and the resulting child is indeed deaf from birth. In the first case, no one would disagree that Alice has caused her child harm. She intentionally and knowingly took a substance during the course of her pregnancy that resulted in her child being born deaf. Alice wronged her child, in that she caused harm that otherwise would not have befallen it.

Carol's case, on the other hand, is where we run into the non-identity problem. Intuitively, Carol has caused her child harm. Well . . . she has, hasn't she? She chose an embryo which was known to have a genetic defect that resulted in her child being born deaf. Yet if we were to analyze every step that Carol took, it's a bit more difficult to pinpoint exactly how and why Carol did her child wrong. If she hadn't chosen the embryo that she did, then that particular child would never have been born in the first place. As for that child's quality of life, could we successfully argue that it has been reduced? Again, this particular child would have been born deaf no matter which candidate embryo had been selected. And even if it had been possible to repair the genetic defect prior to implantation, and this step was never taken, nevertheless, is this child's life as a deaf person so reduced in quality that its life is not worth living?

Deaf people alive today would argue that their lives are indeed worth living. Many would even refer to their disability as being *cultural* in origin, and not physical! This position on disability is not as farfetched as one might think, and it is a contentious issue. In America, for example, a very small minority of PGD clinics have reported the use of their technology "to select an embryo for the presence of a disability" (Camporesi, "Choosing Deafness"). In Great Britain, however, section 14(4) subsection 9(a) of the Human Fertilisation and Embryology Act of 2008 specifically bans the selection and implantation of embryos that are known to have an abnormality that will develop into "a serious physical or mental disability."

Replace "Carol" with "Neolution," and you can see that the creation of the *Leda* and *Castor* clones in *Orphan Black* is an excellent example of the non-identity problem. The clones were intentionally created with the knowledge that they would have a genetic defect that would not only render them infertile, but also eventually lead to a fatal illness. But for this fact, however, none of the clones would ever have been born! *Yet even with this intentionally created, life-ending defect, the clones themselves would argue that their lives are very much worth living.* If that's the case, then how could the clones have been wronged by having been brought into existence? Arguing from the premises to the conclusion, here again is the paradox of the Implausible Conclusion. We are firmly trapped in a philosophical rabbit hole.

Knowledge of Causes

The fact that the clones were purposely created with a genetic defect is revealed during a conversation with one of the original geneticists. Rachel Duncan, a leading executive at the Dyad Institute, and also a self-aware *Leda* clone, asks her adoptive father, Dr. Ethan Duncan, why, of all the clones, only Sarah can bear children. "Why Sarah? Of all of us? How is it the unmonitored tramp was successful?" Dr. Duncan replies, "In her fertility? Huh! Rachel, she's a failure, not a success. *You are all barren by design*" ("Variable and Full of Perturbation").

Drs. Susan and Ethan Duncan had been recruited by the military in 1976 to work on a proof-of-concept project to perfect cloning technology. Susan Duncan's breakthrough on the spindle protein problem made human cloning possible ("To Hound Nature in Her Wanderings"). Ethan Duncan gives Cosima the details. "It was Susan's 'sterility' concept . . . Degrade the endometrium . . . prevent ovarian follicles from maturing." Indeed, "Normal development was the prime directive. This was the least invasive solution. Unfortunately, we didn't, uh, . . . foresee the consequences" ("Things Which Have Never Yet Been Done").

Neolution wants to continue the Duncans' research, but Ethan Duncan doesn't want to help them. Rachel can't understand why. She asks him, "Why would you deny existence to more of us? We're your life's work." Dr. Duncan replies, "Well, since none of us seem to know what you're actually for, I think your attempts to make new clones should meet with failure" ("By Means Which Have Never Yet Been Tried").

The synthetic genetic sequence that ensured that the clones would survive has resulted in them eventually succumbing to a fatal illness. Whereas the *Leda* clones eventually die from a fatal respiratory illness, the *Castor* clones' symptoms are neurological. We see the *Castor* clones being subjected to periodic tests of their reasoning abilities in order to gauge the extent of their illness, and we also see them suffering from seizures. Dr. Virginia Coady reveals to Sarah that she survived the *Castor* prion infection "Because *Castor* and *Leda* have the same disorder" ("Certain Agony of the Battlefield").

The Neolution cabal creates human life—defective clones—as a means to perfecting their eugenics program. They treat the clones with no respect whatsoever for their autonomy or individuality. It's true that they have given life to the clones. Without Neolution's efforts at advancing its eugenics program, not a single one of the clones would have come into existence. But regardless of this fact, because they treat these human beings as little more than lab rats (right down to a patent claim and a unique alphanumeric identifier built right into their DNA), Neolution's actions are not just immoral and harmful. They are absolutely evil.

Governed by Sound Reason

Contemporary philosopher Derek Parfit believed that resolving the paradox of the Implausible Conclusion required the creation of a new theory of ethics, which he labeled "Theory X," but he was unable to create such a theory. Contemporary philosopher David Boonin has in turn taken the non-identity problem and identified the five premises that lead to the paradox of the Implausible Conclusion. Boonin says that to successfully resolve the non-identity problem, we must be able to strike down at least one of the five premises. This would show that the paradox is the result of faulty deductive thinking. In *The Non-Identity Problem and the Ethics of Future People*, Boonin looks at all of the arguments made to strike down each of the five premises, and he finds all of them lacking. For Boonin, we appear to be stuck with the Implausible Conclusion of the non-identity problem.

The non-identity problem has been applied to environmental ethics. Suppose the world today is faced with the choice of continuing to burn coal or transitioning to alternate energy sources. Continuing to burn coal will result in a higher standard of living today; but, due to the increased release of pollutants into the atmosphere, this will mean that people born one hundred years from now would suffer from health problems that would reduce the average lifespan to, say, fifty years. Transitioning to alternate energy sources would result in a relatively lower standard of living today, but people living a hundred years from now would have an average lifespan comparable to what it is today. In the first scenario, it would seem intuitively correct that people today will have wronged people who will be alive one hundred years from now. The actions of people in the present will result in physical harm to future people—a significantly reduced lifespan. Yet the people alive a hundred years from now under the first scenario (continue to burn coal) could not have come into existence in the first place had people today decided to transition to alternate energy sources! And although a reduced lifespan would be viewed as negatively affecting one's quality of life, nevertheless, even these future people would argue that their lives are worth living.

Contemporary philosopher David Benatar has pushed the non-identity problem to its ultimate, and shocking, conclusion. Benatar observes that human life is full of bad things that happen to people. Indeed, many more bad things seem to happen than good things overall. Life is full of grief, heartache, disappointment, and pain. Since it's wrong to harm someone, then the greatest harm that one can inflict on any human being is to allow them to come into existence in the first place!

Benatar argues from the non-identity problem to a position known as antinatalism. Since bringing people into existence automatically condemns them to a life of suffering, regardless of whether they are even aware that they are suffering, and regardless of any optimistic mindset that they may be laboring under, it is wrong to bring more people into the world. The human race should be allowed, indeed encouraged, to die out!

History Yet to Be Written

When a problem is either not clearly stated, or if there are facts that are missing, then a paradox may result. A paradox is a conclusion that appears to be contradictory, even though it has been arrived at by a sequence of logical steps from a set of reasonable premises. To understand paradoxes and how they can be resolved, we can take a look at the appropriately named "twin paradox" of physics.

We take a pair of twins, and send one of them off in a rocket ship, while the other twin remains here on Earth. When the spacefaring twin returns to Earth, we find that the twin that remained on Earth has aged *more* than the twin that went out into space. This doesn't make any sense, until we invoke the time dilation of special relativity. Given the correct perspective, the difference in the twins' ending ages is explained, and the paradox is resolved.

Although many philosophers have weighed in on the resolution of the non-identity problem, a very simple solution has been proposed by philosopher Rivka Weinberg. Weinberg notes that the issue that makes the non-identity problem paradoxical is that it combines metaphysics and morality into a problem that should be strictly moral. In other words, the heart of the non-identity problem is that without the procreative choice having been made, the person in question never would have been born in the first place.

The wording of the non-identity problem makes it seem as though the very fact of someone's existence is some "thing," like a "good," that has been conferred onto someone. But this makes no sense, as the person to whom existence has been "conferred" didn't exist in the first place! Existence isn't a "good," rather, it is a condition, or prerequisite, for all the good and suffering that a person subsequently experiences. Separating the metaphysical issue of existence from the moral issue of whether a person knowingly born into the world with any disadvantage, such as a genetic disability, has been harmed by the procreative choice that was made to create this person means that we can focus on the harm that does befall this person.

It makes no sense to use the metaphysical excuse that, had the embryo with the genetic defect never been chosen, then the resulting person would never have come into existence in the first place. Rather, we should focus on the fact that someone, whether a person making a choice for personal procreative reasons, or an impersonal, evil, corporate entity like the Neolution cabal, has indeed wronged this future person.

A choice was made to knowingly create a person with a disability that would result in a reduced quality of life. Choosing an embryo with a known genetic defect that would reduce the future person's quality of life, *even if overall that particular life would still be worth living*, does indeed harm that individual. It makes this new person a means to some end, and thus denies that person their autonomy and their individuality. It also violates our feeling that the act of procreation invests us with great moral responsibility. We should not casually bring forth a new person into the world without doing everything possible so that this person will face as few disadvantages in life as possible. *In a sense, this moral position makes Sarah (and her now pregnant twin, Helena) no different than the Neolution cabal!* This perspective on the non-identity problem—that it is only a moral problem, and not a metaphysical *and* a moral problem—resolves the paradox of the Implausible Conclusion, and gets us out of the rabbit hole.

At the very end of Season Two, we see the *Leda* clones in Felix's loft apartment, dancing to a reggae tune ("By Means Which Have Never Yet Been Tried"). Although clones, they each have their own style of dancing, of grooving to the rhythm of Life. Their styles of dancing reflect their individual natures as we've come to know them, and the unique way they each approach the problems they face—Sarah with arms and fists held up front and high; Alison reserved and controlled; Cosima fluid and sensuous; Helena spastic and chaotic.

The fact that the *Leda* and *Castor* clones struggle to live their lives on their own terms, despite their genetically engineered illness, and in spite of the constant interference of the Neolution cabal, reveals to us that *Orphan Black*'s position on the non-identity problem is in fact Rivka Weinberg's solution. The conditions that conspired to bring about your existence aren't *the* issue in living. Rather, the issue is living your life the way you decide to live it. Every human life is worth living, *as long as you're able to make a conscious decision to actively fight to make your life worth living*. Thankfully, no matter how grim the future may look—and there will always be such times—we're all hard-wired to see our lives as worth fighting for. With Life, there can really be no other way.

Our lives are fearfully and wonderfully made, by our own two hands, one day at a time.

2
Go Ask Alison

DANIEL MALLOY

The science of cloning is complex, and only becomes more complex when applied to human beings. But, as *Orphan Black* shows, the physical science of human cloning may be the easy bit. Because when you clone a human being, you don't just get a copy of an organism. You get a person. And people are complicated creatures.

Like any experiment, Project Leda demanded observation. To acquire the capacity to observe, the experimenters employed the monitors. The monitor's job was to observe and report on the life of their assigned clone. Some monitors had some information about the nature of the experiment, as with Daniel Rosen and Delphine Cormier; others, like Donnie Hendrix, were completely misled about the nature of their own work.

There's an inherent problem with gathering data for social sciences. Fundamentally, social scientists are people observing and interacting with other people. This presents two obstacles to the objectivity sought by scientists: first, since the observers are themselves people, they bring a wide variety of biases and prejudices with them that shape their observations in subtle and not-at-all subtle ways. Second, since the observers usually also interact with their experimental subjects, there is an ever-present risk that these interactions influence the subjects in ways the observers can neither predict nor account for.

This problem puts data gathering, and the monitors, at the core of a central problem in the philosophy of social sciences. The problem concerns the nature of the social sciences themselves. On the one hand, a tradition espoused by philosopher John Stuart Mill (1806–1873) and social psychologist Émile Durkheim (1858–1917) views the social sciences as no different from natural sciences like physics or astronomy. The purpose of data-gathering in the social sciences, according to this tradition, is the same as in natural sciences: to formulate or test hypotheses. A social scientific theory would then be judged by its ability to generate successful predictions.

A competing tradition, advocated by sociologist Max Weber (1864–1920), argues that since humans are more complex beings than the subjects of other sciences, the prediction-generating paradigm is an inadequate one for the social sciences. Rather than trying to predict human action—a task that may not be possible—this tradition views the purpose of social sciences as interpreting and understanding human behavior.

This divide can also be understood as an external-internal distinction. The prediction paradigm strives to view and grasp human actions from the outside, as events in the world like any other. It seeks the causes of human actions in largely the same way that the natural sciences seek the causes of natural phenomena. The interpretation/understanding paradigm, on the other hand, strives to conceive human action as the actor herself does, from the inside.

Instead of causes, social scientists in Weber's tradition seek motives. Motives are fundamentally different from causes in two ways. First, motives are unique to conscious beings—when Helena stabs someone, she's acting on a motive. The blade she uses is not. But, second, a motive, unlike a cause, does not determine a particular outcome. When Alison lets Aynsley die, for example, she has motives for not saving her; but she also has motives for saving her. Her choice determines which set of motives she acts upon. Causes, however, preclude choice.

So, in studying the clones of Project Leda, the predictive paradigm would call on us to observe and record the sequences of events that define their lives, looking for repeating patterns and attempting to find regularities. Enough repeating patterns and regularities could indicate a cause-and-effect relationship, and thus allow for accurate predictions of future actions. These sorts of observations only require the invasive surveillance of monitors in order to ensure the most complete access to the clone.

The interpretation/understanding paradigm, on the other hand, seeks to grasp what it's actually like to be a clone. For this approach, monitors are necessary because the only way to get an answer to that question is for the clone to share it with someone. That kind of sharing demands some sort of relationship.

The interactions of monitors and clones also raise ethical concerns. An observer may, through access to knowledge undisclosed to the subject, infringe on the subject's ability to make choices. Alison might not have consented to marry Donnie, for example, had she known that he was spying on her.

Donnie's Dilemma

Pity Donnie Hendrix. He fell in love with a clone, married her, and got tricked into spying on her for the organization that spawned her. And things just got worse from there. But how Donnie nearly ruined his life and his marriage isn't really our problem. How his involvement ruins Project Leda's experiment is. There are a couple of problems posed by Donnie's involvement in data gathering.

First, Donnie doesn't know the nature of the experiment. This both helps and hurts his credibility as a data gatherer. If Donnie knew the nature of the experiment, he would have a better idea of what data was relevant and what was not—what to include in his reports and what to leave out. At the same time, however, if Donnie knew the nature of the experiment, that might skew his observations—in looking for data relevant to Project Leda he might, unfortunately, dismiss information that was relevant under the impression that it wasn't.

So, perhaps it's best for Project Leda's data collection that Donnie doesn't know what's going on. Unfortunately, people don't just spy on one another and give information to nameless authorities—well, not often. To secure Donnie's cooperation, Project Leda needed a cover story. Donnie's ignorance might be a different matter if he hadn't also been deceived about the experiment. Donnie believes that his wife is the subject of an ongoing sociological study. As such, his observations are based on that presupposition. He's observing his wife as he believes a sociologist would, looking for the kind of data that would be relevant to sociology. Now, sociology is a pretty broad field, so his observations may indeed be relevant to Project Leda—at least some of the time. But Leda's interest goes far beyond the sociologically relevant.

There's a problem that goes beyond Donnie's observations, though. He's not just a passive observer in Alison's life, after all. He's her husband. As such, he's an active participant in her life—making decisions that affect her in various ways, working with her in myriad ways, and thus influencing the "results" of the experiment. There's no problem with this, insofar as Donnie is just Alison's somewhat schlubby husband. But since he is also her monitor, his involvement in her life runs the risk of skewing the results, for a couple of reasons and in a couple of ways.

First off, Donnie must be aware, on some level, that he is monitoring Alison. That calls into question his motives for certain actions. How many decisions has he made over the years, in part because of the data it would provide the "sociological study"? Did it push him to propose? Or to agree to adopting children? And remember that Alison's (as it turns out, correct) suspicion that Donnie was her monitor led to her infidelity with her neighbor. In retrospect, he got off lightly compared to Aynsley.

But Donnie's motives, and how the experiment may have altered them, are really only a concern for him and Alison. From a scientific perspective, the problem is that Donnie's involvement might skew his observations. For example, suppose he and Alison decide to get a bit adventurous in the bedroom (feel free to picture this if you like; I won't): does he report that? How does he decide its relevance to the sociological study? And does his own desire for privacy influence what he chooses to share with Project Leda?

Making Donnie a monitor wreaks a good deal of havoc on Leda's experiment. His deep involvement in Alison's life makes the data he provides suspect. His ignorance about the experiment jeopardizes the relevance of his reports. And his status as Alison's husband raises basic ethical concerns about the level of manipulation she's undergoing every day.

The Drawbacks of Daniel

Many of the problems surrounding Donnie and Alison's involvement in Project Leda stem from their ignorance of the experiment itself. Donnie doesn't know what he's looking for as a monitor, and Alison doesn't know that she's the subject of an experiment at all. If their ignorance is the problem, then perhaps it's addressed or corrected by the existence of Rachel Duncan and Daniel Rosen.

Both Rachel and Daniel know that the experiment is happening and what their roles are in it. Rachel is the only self-aware clone in Project Leda. Daniel knows both that Rachel is a clone and that his job is to observe and report on the activities of a clone.

Although their relationship is obviously very different from the Hendrixes', their mutual knowledge makes Rachel and Daniel partners in the experiment in a way that the Hendrixes can't be. Rachel, as the self-aware clone, knows that her actions are those of a clone. She knows that Daniel will report on their interactions—hell, she'll get the reports herself. She even has approval over who her monitor is.

Rachel and Daniel's knowledge, however, presents just as much of a problem as Donnie and Alison's ignorance. The problem apparently begins with Rachel's childhood. Rachel was raised as a clone. She was not, however, raised as a child.

It isn't so much the knowledge that she is a clone that is problematic for Rachel, as it is the attitude toward clones (including herself), that was instilled in her.

Rachel thinks of clones as products, property, as prototypes. She and the others are things, not people. Since Rachel thinks of herself as a thing, she views others in the same way. Just consider the way she uses her monitors, Daniel and Paul. They are little more than toys to her. Her sisters are property of the Project, not people with their own thoughts, feelings, hopes, and dreams. Even her "father" is little more than an obstacle in her world. Raised as the object of an experiment, knowing she was the object of an experiment, Rachel simply isn't capable of interacting with people as people.

While that may be an unfortunate side-effect of her childhood (and, some have argued, of cloning), it doesn't in itself invalidate the experiment. What invalidates the experiment is that Rachel is supposed to serve as the control group. In any decently designed experiment, there is a control group that serves as a point of comparison. When testing a new medication, for instance, trials require at least three distinct groups of patients: one that gets the drug, one that gets a placebo, and one that gets no treatment. If the drug's effects are not significantly different from the placebo's, then the drug is considered a failure. Being the self-aware clone, Rachel is, in a sense, the standard against which the other clones are measured. But because she hasn't been raised to consider herself or others like her as people, she will provide a terrible baseline. Comparing data from others to her data is going to skew the results of the entire experiment.

A further issue with this methodology is the need for a further control group. Rachel was raised in an odd way that had profound effects on her. She may present a better face than Helena, but in her own way she is far more damaged. In order to determine whether her status as a clone or her unique childhood—or some combination of them—is responsible for her cold approach to the world, further controls would be necessary. Would a naturally conceived child turn out like Rachel if she were raised believing herself to be a clone and the object of an experiment? I conjecture that she would, but since there's no data, that is only a supposition.

Together, Rachel and Daniel represent the extremes and pitfalls of the prediction model of the social sciences. When we study human beings in the same way that we study mere things, with no consideration for their internal lives—their motives, hopes, dreams, and fears—we run the risk of reducing them to things. As human beings, as persons, participants in social science experiments are entitled to a certain degree of respect. Included in that respect is a duty to get their fully informed consent to participate in the experiment to begin with. Rachel and Daniel, and indeed the whole of Project Leda, have no problem violating this basic tenet of ethical experimentation. Even when it seems that they are willing to acquiesce to these basic rules, the honchos of Project Leda, including Rachel, push beyond all reasonable bounds. Recall when Leda was forced into the light and offered a more open and co-operative deal to the clones: it turned out that they were in actuality trying to cement their grasp by obtaining the clones' agreement that their genetic material—including the genetic material of any and all offspring—was the exclusive property of the Project.

Delphine's Difficulty

Perhaps the best monitor would be somewhere between Donnie and Daniel. Donnie is too involved in Alison's life and ignorant of the nature of the experiment; Daniel, on the other hand, is too detached and uncaring about the clones, treating them as things rather than as people. There is a monitor who falls between those extremes: Delphine Cormier. Delphine has full knowledge of the experiment like Daniel but also cares about her assigned clone, like Donnie.

Unlike the other monitors, Delphine is actually a scientist, and therefore more qualified than any of the other observers to take part in the experiment. Delphine's background in immunology rides the line between the natural and social sciences, making her in some ways the ideal monitor. She can keep track of the physiological and biological condition of her assigned clone while monitoring and reporting on her behavior. She may even, unlike other monitors, have some inkling about what behaviors may and may not be influenced by the clone's status as a clone. Further, a background in medicine discourages the sort of objectification of the experiment's subject that Rachel and Daniel fall into. On paper, Delphine's background and skill set seem to make her the ideal monitor. She can get just involved enough in her clone's life to keep careful watch over her while maintaining enough distance to avoid inadvertently manipulating her actions and thereby skewing the data.

Unfortunately, Delphine is not allowed to pursue her monitoring duties unencumbered. Influenced by the threat of the Proletheans, Leekie pushes her to get closer to Cosima than she really should. This in turn leads to the very real romantic relationship between Delphine and Cosima, which ultimately leads to the immunologist having some of the same epistemic weaknesses as a monitor as Donnie Hendrix: she reports selectively. The only difference is that her selection is based more on her feelings for Cosima than the needs of the project or any potential embarrassment the information may cause her. Regardless, however, it still skews her data collection. To give just one example of how her feelings compromise her objectivity, Delphine only reports that Cosima knows she's a clone and is in contact with other clones in order to help Cosima.

Further, perhaps oddly, Delphine's relationship with Cosima leads to her being ethically compromised as well. She manipulates the clone as much, if not more, after revealing herself as a monitor—only now, it is in what she believes to be Cosima's own best interests. Before the reveal, she introduced Cosima to Leekie without revealing their relationship or Leekie's true interest in Cosima. Either of those pieces of information would have altered how Cosima behaved if she had known them.

Drawing Conclusions

Projects Leda and Castor comprise the ultimate twin study. As with any twin study, you begin with a pool of subjects whose genetic code is as close as possible to identical. Place those genetically identical subjects in different circumstances and, in theory, you can observe and measure the influence that environment has on personal development. In theory, had Sarah or Cosima or Tony been raised by Connie, they would have become soccer moms like Alison. And the other way around, had Alison been raised by Mrs. S, she would have become a tough, self-reliant thief like Sarah. Part of the point of the monitors is to determine whether that hypothesis holds true.

But, because of the flaws in Project Leda's data-gathering method, the monitors' reports neither support nor undermine the hypothesis. The design of the experiment is so inherently flawed as to be meaningless. No conclusions whatsoever can be drawn, no predictions can be made, based on this experiment. The control group is a mess, the selection of data gatherers is seemingly random, the quality of their reports no doubt varies wildly, and there's little to no guarantee than the reports are at all accurate.

But what if we consider the experiment from the interpretive/understanding approach, rather than the predictive one? Do the monitors' reports enable us to understand the inner lives of the clones?

Not really. Recall Donnie's reports: in spite of being as close to his clone as any monitor, as involved in her life as is humanly possible for any two people, Donnie never even suspected Alison's motives for many of her actions. She managed to keep the Clone Club, and her later infidelity, completely secret from him. Donnie's reports, from what we saw of them, were generally superficial at best. He merely told Leekie what his wife had done—he never even speculated regarding her motivations.

From any perspective, the Leda experiment, considered as an experiment in social science, is a complete and utter failure. Its only accomplishment was to wreck the lives of everyone unfortunate enough to come into contact with it—monitors and clones alike.

3
When Clone Club Looks for Answers

JOHANNA WOLFERT AND ADAM BARKMAN

If someone asked you to tell them what *Orphan Black* is all about, we'd guess your response would have something to do with its characters trying to find the truth. The show follows Sarah and her sisters as they try to figure out their place in the ever-expanding conspiracy of cloning and subterfuge. Their quest for answers is central to *Orphan Black's* storyline, and though Sarah and the other women are genetically identical, the roles they end up playing in Clone Club are dramatically different.

The Dreadlock Science Geek

One of the first clones we meet is Cosima Niehaus, a cheeky girl distinguished by her glasses and ponytail of dreads. She can also be identified as the one wearing a lab coat or staring down a microscope—she's a postgraduate student in evolutionary developmental biology, which she fondly refers to as evo-devo.

A somewhat quirky scientist, Cosima's passion for biology often finds her "totally dorking out" ("Entangled Bank") over her various experiments and projects. Most of these projects are focused on her ongoing study of the clones. When she introduces herself to Sarah, Cosima is at university working toward a dissertation on epigenetic influence on clone cells, but it isn't long before she starts doing full-time research at the Dyad Institute, trying to find out everything she can about her own biology.

The Science which Considers Truth

Way back in the early days of science, before the periodic table or microscopes or the discovery of DNA, there was another man with a passion for biology. Like Cosima, he spent many years of his young adult life in school. After he left, he devoted his time to research, including researching the natural world. However, this man's study of science or "natural philosophy"—impressive as it was for the time—was eclipsed by his study of philosophy proper or "the science which considers truth." This man turned out to be one of the most important if not the most important philosopher ever: Aristotle.

Aristotle believed that the ultimate goal for humans was to be happy. He reasoned that because humans are rational creatures, our happiness is going to involve rational things. A flower can flourish with sunlight and a nice patch of dirt. A sheep will lead an excellent sheep life in a lush field of grass, living with its herd under the watchful eye of the shepherd. We're going to need food and protection too, but we—our rational souls—also need more. For us, a large part of our happiness is found when we seek truth. Aristotle was convinced that knowing and understanding what the truth is—for its own sake, not just as a means to some other end—is essential for humans.

We Need to Know

Like fellow biologist Aristotle, Cosima is also passionate about seeking truth. In her case, this means finding answers to the plethora of questions that come with being a clone, and she believes that the way to get there is through science. When Sarah is first inducted into Clone Club in "Variation under Nature," the newcomer is desperate to understand what's going on. Cosima is happy to oblige—but only on the condition that Sarah give her the briefcase she got from a German clone. Sarah, naturally, wants her information first, and the two remain locked in a stalemate until Cosima points out that "the answers are in the briefcase." It's the scientific contents of the briefcase, which include syringes of blood and other genetic samples, that Cosima believes will get them to the truth of their origins. Her conviction that this matters is obvious as she tries to convince Sarah to help her on her quest for truth: "Who's the original? Who created us? Who's killing us? We need to know."

From the moment we first meet her, Cosima only becomes more deeply invested in the biology of the clones. She can't resist a new opportunity to get closer to the truth, even when it's dangerous. Take her attitude toward Helena, for example. While most of us would agree that a religious fanatic who just killed one of your sisters is typically someone to be avoided—especially when your name is up next on her hit list—Cosima has the opposite idea: "If she's not dead, we need to find her and find out what she knows . . . We need to find out who she is, Sarah; she found us, she's got answers" ("Effects of External Conditions").

Cosima's idea of useful information sources isn't limited to violently unstable assassins. Unlike most of Clone Club, she doesn't seem to mind when monitors become an issue ("Parts Developed in an Unusual Manner"). Though we can't deny that her attraction to her new monitor was a factor, it's also in her insatiable curiosity that Cosima ignores Sarah's emphatic warnings to stay far away from Delphine. Instead, she turns around and urges Sarah not to ditch her own monitor, a sinister ex-military badass, arguing that they could use him to "finally get some answers."

The passionate biologist is even willing to strike into the heart of enemy territory to get closer to the truth behind their existence. In "Endless Forms Most Beautiful," Dr. Leekie of the Dyad Institute presents contracts to the various members of Clone Club, doing his best to convince them to sign the agreement. To the others, he brings tantalizing proposals of safety and privacy; he promises that the people of Dyad will get out of their lives. When he approaches Cosima, though, he extends the opposite invitation: "This is your complete, sequenced genome—3.2 billion base pairs. My offer is the freedom to study yourself and your sisters—unfettered research." He knows that it's a chance to dig deeper that will tempt Cosima the most, and it works: even though she was the one who discovered the patent in their DNA and warned that "any freedom they promise is bullshit," she can't resist Leekie's offer and is soon designing her own lab in the heart of Dyad headquarters. Despite all the risks and dangers of the job, it is imperative for Cosima to seize this golden opportunity to get answers.

Even when It Hurts

Cosima's dedication to truth is particularly evident when things get rough. The mysterious and severe respiratory illness of their German sister was always a looming threat to Clone Club, but it becomes a lot more real to Cosima when she starts coughing up blood in "Endless Forms Most Beautiful." However, getting sick doesn't stop her—if anything, it only makes her more determined. In "Mingling Its Own Nature with It," she watches hours of painful video logs from another clone who died of the same illness; she even performs an autopsy with her fellow scientist and girlfriend, Delphine. Cosima is clearly shaken by the sight of a dead body identical to her own, hair fallen out and internal organs ravaged by disease. But when Delphine gently suggests that it might be too much for her, Cosima brushes her off, annoyed: "Of course I can handle it. Don't be a bitch." She doesn't let her failing body slow her down, continuing her research even when she has to drag an oxygen tank behind her in the lab.

The way forward isn't always clear, either. Cosima's the first to admit that there often aren't any answers in sight, confessing that "None of us have any idea what we're doing, we're just poking at things with sticks" ("To Hound Nature in Her Wanderings"). And even when a lead does pop up, sometimes all her hard work just seems to be proven futile in the end anyway—like when the original genome that's critical to finding her cure turns out to have been destroyed in a lab fire ("Ipsa Scientia Potestas Est").

In spite of all the threats and challenges thrown her way, Cosima remains devoted to her quest for truth, insisting, "We have to know our own biology; that's what this is all about" ("Endless Forms Most Beautiful").

The Soccer Mom

The other member of Clone Club that we meet right off the bat is Alison Hendrix. Though she and Cosima have a lot in common biologically—like a few billion identical base pairs in their DNA, to start—and share a face, Alison is pretty much the polar opposite of her sister. Before becoming self-aware and being inducted into Clone Club, she led a spectacularly ordinary life: after graduating from high school as valedictorian, she went to college to study kinesiology. There, she met Donnie, and the two eventually got married and moved to the suburbs to settle down with their two adopted kids. Apart from the occasional period of part-time work at her mother's shop, Alison devotes the majority of her time and energy to her duties as the quintessential soccer mom.

Happy, Safe, and Sound

Now, if Alison were to find a kindred spirit in Ancient Greece, our first guess would probably be Epicurus. He was another philosopher who lived around the same time as Aristotle, although his work didn't really end up taking off in the same way. Like Aristotle, Epicurus believed that happiness is the ultimate goal for humans. However, his reasoning from there went in a very different direction.

Epicurus argued that happiness is the absence of suffering, writing that "The limit of pleasure is the removal of all pain," which is the third of his *Principle Doctrines*. For Epicurus, a major part of avoiding pain was ensuring your personal safety: "In order to obtain protection from other men, any means for attaining this end is a natural good" (*Principle Doctrine* 6), and "Those who possess the power to defend themselves against threats by their neighbors, being thus in possession of the surest guarantee of security, live the most pleasant life with one another" (*Principle Doctrine* 40). Epicurus didn't have all that much to say about truth—in the grand scheme of things, it's more or less insignificant, and you definitely don't need it to be happy.

Avoiding the C-Word

Seeking the truth isn't exactly a priority for Alison, either. On the contrary, she makes it clear from the get-go that she would prefer to stay as far away as possible from the whole clone business—"Why me? I never wanted any part of this," she laments during her first appearance in "Instinct." Rather than finding the truth, Alison's goal is the happiness that comes from not suffering—in her case, this means that her family isn't threatened and that anyone interested in her as a clone stays far away. She also doesn't see the need for anyone else to look for answers, preferring everything to stay on a need-to-know basis. When Sarah seeks her out for the first time, Alison refuses to tell her anything about the situation, saying that it isn't her responsibility. Instead, she becomes furious that Sarah showed up in her territory: "This is my neighborhood, you wait for a call" ("Instinct").

Of all the members of Clone Club, Alison is the least involved in the higher levels of their creators' conspiracies. She never takes the initiative to poke around for answers and wants nothing to do with her sisters' schemes: "I don't want to know anything, leave me out of it . . . plausible deniability" ("Nature Under Constraint and Vexed"). Sure, she'll dress up like one of her sisters under desperate circumstances—protecting Sarah's relationship with Kira in "Effects of External Conditions," misleading Dyad investigators in "The Weight of This Combination"—and she even agrees to take her turn fostering Helena ("Ruthless in Purpose, and Insidious in

Method"). However, she's not actively involved in Dyad; we never see her helping decode Professor Duncan's book or hunting down people who might have more information. The closest she gets is chucking Clone Club some money in case of sketchy transactions.

Housekeeping

In stark contrast to Cosima's relentless curiosity, Alison has no desire to dig deeper. As we watch her sisters toe the line between life and death, Alison's domestic struggles feel like part of a different world. While Helena is being artificially impregnated by the Proletheans, her suburban sister's life remains wrapped up in her play—the worst place Alison gets carted off to is rehab.

When Sarah and Helena wind up as prisoners in a desert military base later on, the soccer mom is still blissfully disconnected from it all, focused on her candidacy for school board trustee and blossoming drug business. Alison's perfectly content to leave the ominous shadows for the rest of Clone Club to deal with; her attitude is summed up pretty well in "The Weight of This Combination," where she hangs up on Sarah—who's trying to warn her about the looming threat of Castor—to yell at the ref during her kid's soccer game.

When Dr. Leekie stops by Alison's suburb with a contract in "Endless Forms Most Beautiful," what he offers her is protection. The shrewd negotiator knows that dangling the potential to get to the truth isn't going to appeal to the devoted mother like it would to her science-minded sister. In fact, the mention of answers is enough to send Alison into a rage: "I don't want answers anymore! I want my life back! I want my family back, I want my privacy back, I want you out of my life and I want things to be normal again." Instead, Leekie tells her what she's been desperate to hear—that her family will be protected, that she won't have a monitor and that her life will be pretty much free of Dyad. Alison knows that Leekie and Dyad are far from trustworthy, but the promise of privacy and security is too much for her to resist. Barely a few hours later, she signs the contract and faxes it in.

Who's the Spy?

Alison seems to become extremely interested in truth when monitors become a factor. She's obsessed with exposing whoever's been spying on her, and she'll stop at nothing to hunt them down. The first person she suspects is her husband, Donnie; once the thought takes hold, it doesn't take long for her to go completely off the rails ("Effects of External Conditions," "Conditions of Existence," "Variations Under Domestication").

She frantically searches her house and her husband's drawers, and then buys a nanny-cam for her room to uncover evidence of being monitored. Even though it only shows Donnie getting up in the middle of the night, it's enough to convince Alison that he's performing medical examinations on her while she sleeps. She knocks him unconscious with a golf club, ties him to a chair and tortures him with a glue gun to try to make him confess—but he won't, and Alison is devastated by having messed up her family.

Just a few short minutes later, Alison becomes equally convinced that her monitor is actually her best friend Aynsley. Never mind that it doesn't really make sense—it was a simple suggestion from Felix, who'd never even met Aynsley before his afternoon as Alison's emergency bartender in "Variations Under Domestication"—once she latches on to the idea, there's no stopping her. Alison flips immediately to cold anger and barely veiled suspicion, but when the only response from her former confidant is confusion, Alison goes into full attack mode. She sleeps with Aynsley's husband in his car in the middle of the parking lot after figure skating, which leads to a no-holds-barred catfight with the "sleazy watcher spy" in the middle of their street ("Entangled Bank"). Her paranoia of the monitors runs so deep that she chooses to stand by and watch as Aynsley gets strangled to death by her own scarf ("Endless Forms Most Beautiful").

Even though Alison will stop at nothing to find the truth about her monitor, she doesn't want to know this truth for its own sake. Matching with Epicurus's view, the goal is to restore her privacy and personal safety. Her witch hunt is ultimately motivated by her fear that those controlling the experiment will send someone after her. After her violent scuffle with Aynsley, she's terrified to return to her suburb alone and insists that Felix accompany her: "I can't go home, Felix. I assaulted my monitor. I'm going to get black-bagged like Sarah and carted off to who-knows-where" ("Unconscious Selection"). Alison's interest in the monitors, however strong it may be, ends with her escape from whoever's prying into her life.

You Can Run but You Can't Hide

Though Alison makes every effort to pretend the whole clone situation doesn't exist, this doesn't exactly work out for her. She knows that living in denial can't go on forever: "I don't want to divorce Donnie, but I can't keep lying to him. I hate lying to my kids. How can we possibly sustain this insanity?" ("Entangled Bank"). To take the pressure off, Alison turns to drinking and illegal prescription drugs; she's rarely seen without a bottle of pills or a mini flask of alcohol throughout the first season. Her early attempts to cope with her constant paranoia literally push her over the edge—she OD's during her beloved musical and falls off the stage ("Mingling Its Own Nature With It"), waking up in rehab with a broken collarbone ("Governed as It Were by Chance").

It's during her week in rehab that Donnie accidentally finds out about the clones when he stumbles upon his wife talking with Sarah ("Knowledge of Causes, and Secret Motion of Things"). Naturally, once the truth about clones and monitors is out in the open—Donnie thought he was just continuing a sociology study from college—their marriage improves rapidly. Alison also kicks her drinking habits and ditches her happy pills; after all, she doesn't have to live in hiding anymore.

By the end of Season Three, Alison is well aware that she owes her domestic tranquility to the hard work of the rest of Clone Club. Suburban life as she knows it wouldn't be possible without her sisters handling all the clone business behind the scenes, and she's very grateful: "Thank you for protecting us, for making us feel normal. I know that whatever comes next, we'll face it together, as a family" ("History Yet to Be Written").

Not Even Close to the Same Girl

It's easy to see how the similarities between the clones include identical DNA and not much else. One ended up in the lab, one ended up in the suburbs—it's remarkable that they're able to progress from calling each other "bitch" to calling each other sisters.

Just because they're on the same team doesn't mean they're going to start collaborating on projects, however. Even though she mostly stops running from the C-word as time goes on, Alison's priorities are still Epicurean. She has no wish to get involved in the whole looking for answers thing; taking care of her personal and family matters is much more important.

Meanwhile, Cosima continues to seek out answers the best way she knows how: by "following the science" ("Unconscious Selection"). Like her ancient predecessor Aristotle, the modern biologist believes that this search is critically important for its own sake, regardless of the obstacles along the way, and her hunger for the truth keeps drawing her deeper. In many ways, the show's producers are counting on us to follow our own desires for truth. If we weren't just as interested as Cosima in getting our questions about Clone Club answered, *Orphan Black* probably wouldn't be around for very long.

4
Who Owns Clones?

Rod Carveth

The term "cloning" often conjures up images of evil scientists working in secret laboratories engaging in Nazi-like experiments to create human-like creatures for nefarious purposes. As scholar Jason Eberl has noted, images in popular culture at worst, reinforce those images, and, at best, ignore important social and philosophical issues brought about by these portrayals. Unlike the typical media portrayals that Eberl talks about, Orphan Black goes well beyond the stereotypes about cloning, and does address important ethical issues about the process and practice of cloning. One such important issue is: can a corporation own a clone?

Clones are the central subject matter of *Orphan Black*. The term "Orphan Black" refers to orphans (clones) being "in the black," or hidden during the restrictive regime of former British Prime Minister Margaret Thatcher. Sarah Manning, the central character of the show, is a naturally genetically created clone. That is an important description—natural genetic creation—as it has implications for a later discussion.

Just the Facts

When the series opens, Sarah watches a woman who looks exactly like her (except that she's professionally dressed and impeccably coiffed), walk off a subway platform in front of an oncoming train. Then Sarah does what her life of petty crime and failed relationships has taught her to do—she steals the woman's bag and leaves the station. Slowly, Sarah assumes Beth's life, at least as long as it will take to spend all her money.

Soon, though, Sarah learns that she's a clone, with a potentially unlimited number of identical sisters, all of whom turn out to be quite different from each other. Alison Hendrix is a high-strung suburban perfectionist mom. Cosima Niehaus is a scientist studying experimental evolutionary developmental biology. Beth, the woman who killed herself, was a cop. Eventually, Sarah will encounter Helena, a disturbed product of a fundamentalist clone-hating cult.

The "original" (source of genetic material) for Sarah is Kendall Malone. Siobhan Sadler's mother. Ironically, Siobhan ("Mrs. S.") became Sarah's foster mother after Sarah was smuggled through the pipeline, to avoid being captured by Neolution and subjected to medical experiments. Neolution is an organization dedicated to furthering the betterment of mankind's evolution. Their aim is to expand humankind's potential with science (oddly, they seem to think the addition of tails will further this goal). Siobhan explains that she and some of her friends, such as Carlton, would smuggle in children to hide them and Sarah was one of them.

Kendall Malone has had a rocky relationship with Siobhan. For instance, she may have murdered Siobhan's husband, John. Malone was imprisoned in the 1970s and was held at the prison that Ethan Duncan went to, searching for viable donors for his cloning experiments. Duncan was working for a project that was initially a military operation, before the Dyad Institute took over. Duncan and his wife Susan were part of a group of implantation teams believed to be the clones' creators.

While Duncan found a viable donor in Malone, he got more than he anticipated when he learned that Malone had "consumed a twin" in the womb. "Consuming a twin" (also known as "vanishing twin syndrome") is a rare occurrence where twins are in development in the womb and one twin dies in the fetal stage. Rather than the dead tissue being expelled, the second twin absorbs it. This is what happened to Kendall Malone when she was in the womb—she was the living fetus, and absorbed the tissue of her deceased twin.

Because Kendall absorbed her twin, she has two cell lines. That means she has two completely different sets of structured cells within her DNA, in contrast to just about all other people who have just one. These cells will contain different DNA because the twins were dizyotic (came from two zygotes). Thus, when you isolate particular strands of Kendall's genome and separate the cell lines you get what results in the clones that comprise Project Leda and Project Castor ("History Yet to be Written").

Because Sarah came from Kendall's DNA, not only is Kendall Siobhan's mother, Sarah and Mrs. S. are indeed blood-related—in fact, they are genetic half-sisters. Complicating things even further is that fact that Sarah has an identical twin, Helena (who also is Siobhan's genetic half-sister).

About the time that Duncan was succeeding in his cloning trials, Neolution discovered what he was trying to do. They sought to make sure the company was not only a part of this transformative scientific development, but that they also controlled how it evolved, as human cloning could drive a lot of activities in which Neolution wanted to engage.

Duncan eventually found out that people from Neolution had infiltrated every aspect of the project, from the male version of the clones on Project Castor in the military, to the female side with the Dyad. Once Neolution got in, they genetically created infertile clones from Kendall's DNA. Unbeknownst to Neolution, Duncan and his wife succeeded in creating Sarah and Helena, untouched by Neolution's attempt to change evolution. Because these clones were so pure, Duncan had requested Kendall's help to find a good place for them, and Kendall put Sarah with the only person she thought deserved her: Mrs. S.

A few years later Duncan returned to the prison where Kendall had been and told her that Neolution poisoned the experiments once they had successfully cloned her DNA. Duncan revealed that there was a clone (which actually turned out to be two clones, Sarah and Helena) that made it out unscathed. In addition, Duncan told her that Dyad was not able to reproduce any more clones—though Charlotte Bowles was cloned after more than four hundred attempts. Kendall is eventually taken to Iceland by Siobhan and Sarah.

The clones whose DNA was genetically altered was implanted in various mothers. Some of the clones' mothers desired children enough to have opted to undergo in-vitro fertilization, like Alison, Cosima, and Beth's parents. Other mothers may have been approached by Dyad scientists posing as couples who wanted children of their own but needed a surrogate, like Sarah and Helena's apparent birth mother, Amelia.

The female clones were not made aware of *being* clones, so monitors were assigned to them. (Not being self-aware was in their best interest as six self-aware female clones in the Helsinki area were executed.) These monitors were friends or significant others who kept tabs on them and reported on their conditions to the Dyad Institute. The monitors themselves are mostly unaware of the real reasons why they are observing the subjects. Thus, the study of the clones is "double-blind," so that the clones could make their own choices.

The Dyad Institute has the monitors report to research supervisors. Sometimes this happens directly, and sometimes it happens indirectly. For example, Alison's monitor, Donnie Hendrix, reported directly to Dr, Leekie, the then head of research at the Dyad Institute. Paul Dierden, Beth's monitor, reported indirectly to Olivier Duval, who answered to Leekie. The research into the female clones is generally passive (such as observations by the monitors) to avoid making the clones self-aware, though medical tests are conducted on occasion. The goal of the medical research is to find a cure for an auto-immune disorder that attacks the female clones' epithelial tissues, resulting in a terminal condition characterized by violent fits of coughing up blood and a general feeling of weakness. The disorder affected several of the clones, including Katya, Cosima, and Jennifer. Jennifer succumbed to the disease.

The modification of the remaining clones left sequencing on their DNA with certain encoded, encrypted messages which translate into the clones' tag number. These messages are used to tell them apart, and serve as the patent for their creation. Cosima is able to decipher the clones' genetic code and discover a message written in nucleotides: "This organism and derivative genetic material is restricted intellectual property." The message constitutes Dyad's patent. What it implies is that Dyad has patented a modified sequence of DNA already found in the clones' genome. The original clone material was "contaminated"—likely by cDNA. (There is also a possibility that Dyad created the entire clone genome synthetically, a process which would be patentable, and really would cause ethical and existential problems for Sarah and Helena. That, however, does not appear to be the case.) For clones other than Sarah and Helena, there are a number of oddities. For example, Cosima shouldn't even be able to study her own genome without infringing on Dyad's rights ("Endless Forms Most Beautiful").

Patently Absurd?

One of the themes running through *Orphan Black* is: Who owns the clones? This question has serious implications for the characters in *Orphan Black,* including, but not limited to, Sarah and Kira. A person who is considered property is really a slave—a status civilized countries dispensed with in the last century and a half. The situation we encounter in *Orphan Black* raises important philosophical questions. When someone else has the power to regulate what you can and cannot do with your body, what does that mean? How do you view yourself as a result? How are people expected to view you from the outside?

The patent system was created more than two centuries ago with a dual purpose. One is to offer temporary financial incentives for those at the ground floor of innovative products like the combustible engine and the X-ray machine. Patents provide a financial incentive to create. The second is to ensure one company does not hold a lifetime monopoly that might discourage competition and consumer affordability. All patent submissions rely on a complex reading of applicable laws, distinguishing between abstract ideas and principles, and more tangible scientific discoveries and principles. The invention must be "useful, novel," and "nonobvious" and carry a description that enables someone to use it for the stated purpose, according to US patent law.

In the past thirty-one years, twenty percent of the human genome has been protected under US patents. But, now that all 20,000–25,000 human genes have been mapped and sequenced through the Human Genome Project, they are in the public domain, meaning they would no longer be considered "new" for the purposes of patents. Any patents on human genes must specify a new use, such as a diagnostic test. The patentability of such tests gives companies an incentive to create them. A good use of such a patent was for the insulin gene, which led to recombinant human insulin, and resulted in new treatments for diabetes.

Nearly three years ago, it looked as though the matter of genes and patents would be settled. On June 13th 2013 the Supreme Court had an opportunity to make a definitive decision on the issue of whether or not human genes could be patented. The case involved Myriad Genetics Corporation, a molecular diagnostic company located in Salt Lake City. In the early 1990s, Myriad Genetics was involved in the discovery of BRCA1, a gene linked to breast cancer, and BRCA2, a gene linked to ovarian cancer. Myriad was able to clone and re-sequence both genes, and, by 1998, was able to obtain patents on the BRCA1 and BRCA2 genes.

Doing so allowed Myriad to develop BRCA Analysis, a predictive medicine product that could assess the risk of coming down with breast and ovarian cancers. Because Myriad held the patent to the two genes, the company was able to have a monopoly to the most effective means of predicting those two types of cancers in women. A number of groups sued Myriad, including the Association for Molecular Pathology. A legal case, *Association for Molecular Pathology v. Myriad Genetics,* began in the lower federal courts in 2010, and, after a series of conflicting rulings, wound up in the Supreme Court in 2013.

Myriad Genetics claimed in the case that because their discovery of the BRAC1 and BRCA2 genes was like finding a genetic needle in a haystack among our 20,000 human genes, the company should be able to patent them. (Discovering mutations in these genes is what led Angelina Jolie to undergo a double-mastectomy.)

The Supreme Court ruled unanimously that Myriad could not patent genes, because genes were "naturally occurring DNA segments." Like all patents, patents on genes expire twenty years after the date of application, meaning the plethora of existing patents on the human genes themselves will run out fairly quickly. In the case of BRCA1 and BRCA2, that would be in 2018.

But in something of a compromise, all nine justices said while the naturally occurring isolated biological material itself is not patentable, a synthetic version of the gene material may be patented. The Supreme Court has long allowed patent protection for the creation of a new process or use for natural products. Whether "isolating" or "extracting" genes themselves qualifies for such protection became the central argument.

The court also ruled, however, that the creation of "products" through the manipulation of a gene could be legal. Thus, the justices took the position that DNA itself is not patentable but so-called complementary DNA (or "cDNA") can be. Complementary DNA is artificially synthesized from the genetic template, and engineered to produce gene clones. Use of this protein-isolating procedure, known as "tagging," is especially important in mapping and cataloguing the vast human genome.

Patenting Clones

The decision leaves open whether, like in *Orphan Black,* a clone could be patented. Part of that answer lies in the science of cloning. Creating a clone actually isn't just science fiction, but something that actually happens in nature. For example, a few species of jellyfish have a larval stage of their lives where all they do is continually clone themselves. Indeed, any animal that reproduces asexually is a clone.

Humans don't reproduce asexually, so to create a Sarah Manning from whomever the original was, you need a donor egg. After removing the nucleus from the egg, scientists would then fuse another cell's nucleus from the organism they wish to clone into the empty egg (along with a very small jolt of electricity). The now complete egg is then inserted into a surrogate womb where the eggs start to divide. Once the embryo comes to term, a clone is born.

To date, no human clones have been attempted, let alone successfully produced, though the this has not stopped controversy about the process. But there certainly are human clones among us. It has nothing to do with genetic engineering, but, indeed, are products of nature—identical twins. Identical twins are natural clones.

The only real difference between clones and identical twins is the process that creates them. Both identical twins and clones may share the same genetic material, but identical twins come from a natural splitting of a fertilized egg (zygote) inside a womb. (So when Helena looks at Sarah, she sees a clone of a clone.) But even though identical twins are the natural equivalent of human clones, that doesn't mean there aren't small differences. Identical twins are actually more similar to each other than to an artificial clone of the same "original."

Identical twins have the benefit of developing at nearly the same time in the same environment and usually being born to the same parents. This reduces the number of variables that can creep into a child's development. Everything from a mother's microbial makeup to the environment a child is born into can affect the way that child's genes are expressed, clone or not.

Twins are one of nature's great experiments. There have been identical twins who have been separated at birth and raised in different families. So, while they were born with the same genes, they were raised in different environments. In an ethical manner (unlike what Dyad did with the clones), scientists have tracked the lives of identical twins to uncover just how much genetics factor into the course of their (and by extension our) lives. From tracking twins, scientists now know that some human attributes are largely genetic. If one identical twin is tall, it's almost a certainty that the other twin will also be tall. The same is true for some diseases. If one twin develops Alzheimer's, the other one likely will, too.

Tracking twins has also allowed scientists to discover that even though identical twins may share eighty percent of things in common, biology is not destiny. Besides, DNA doesn't copy perfectly during fetal development, imprinting different mutations between clones and between identical twins.

One of the more intriguing genetic anomalies in *Orphan Black* is the fact that Sarah and Helena are "mirror-image" twins. When a fertilized egg splits off into two at a later stage in embryonic timing than normal, there's a chance that once the twins are born, one will have reversed symmetry. Handedness could flip, as could dental structure or even organ placement. Called *situs inversus*, the flipping of organ placement to the other side of the body is what Helena has with respect to her mirror-image twin Sarah.

In addition, when a cell from an organism to be cloned is inserted into a donor egg, the reprogramming that occurs at a genetic level also leaves its stamp on the clone that will differ slightly from the original. In short, clones and identical twins are defined by their genetic similarity, but the similarity is never one hundred percent.

Genetic engineering differs from cloning in that whereas cloning produces genetically exact copies of organisms, genetic engineering refers to techniques in which scientists manipulate genes to create purposefully different versions of organisms. Thus, though the clones in *Orphan Black* are derived from the same genetic material (Kendall Malone), Malone's genetic material was genetically altered (to make the clones sterile, for example). This genetic alteration may be why some of the females may have a disease that affects their lungs and gives them serious coughing fits, and why the males suffer from seizures.

So, because the clones have had at least some genetic engineering done on them, the genetic sequences involved are not technically "naturally occurring," which is what the Supreme Court decision in the Myriad case turned on. As a result, controlling the clones through ownership of their DNA may be legal and very well could be held constitutional by the Supreme Court. So, it is entirely possible that the modified DNA in the Dyad Institute's clones would fall under this exception and Dyad could, indeed, own the clones. If this is the case, Dyad's patent could currently be enforced under the most recent Supreme Court decision. So, even though *Orphan Black* is fiction, the legal status of the cloning technology mirrors current US law.

The Ethics of Cloning

Does the Dyad Institute own the members of the Clone Club? If, under the law, it does, does that entail that's the way that it *should* be? Is ownership of clones *ethical*?

If clones are viewed as possessions rather than as people, it would, in many ways, be a tremendous boon for science. Certain medical pursuits might be easier, because one might think that the owners of the clones would be able to study them without incident. Diseases could be clearly identified because their genetic and environmental influences could be controlled. Thinking of clones as property has other benefits as well. Some of those benefits are explored in *Orphan Black* (consider the use of the Castor clones as ideal soldiers).

The most obvious moral argument against patenting genetic sequences is that genetic sequences are, taken together, critical components of concrete, individual, human beings. These human beings have autonomy that we have a moral obligation to respect. Regardless of the way that these human beings are produced, they still have all of the same rights as human beings that were created in more traditional ways. In much the same way that it is immoral to have slaves, it is immoral to assert ownership claims over real, concrete human beings. If we think of human clones as possessions, we encounter other moral issues as well, familiar from the days of slavery, such as the status of clone marriages and the status of the offspring of the clones. Would clone owners have a right to insist upon or dissolve marriages? Do clone owners also own the offspring of the clones?

In this case, we must weigh the value of preserving intellectual property rights against the rights violations of the clones. People have ownership over the products of their labor, but not if those products are *people*. Humans always rightfully *own themselves*—they cannot morally be owned by others. If the story of *Orphan Black* were to play out in real life, the Supreme Court would need to revisit the decision that it came to in the Myriad case. It is interesting to think about the ways in which they might draw a distinction between genetic material that is not naturally occurring in the Myriad case and material that is not naturally occurring as parts of Leda and Castor.

Orphan Black has created a world where genetic engineering and patents are not merely scientific concepts, but important flesh-and-blood concerns. The series examines the nature of individuality and the relative strength of genes and environment in shaping who we become, while also considering the possible contingencies of our new biotech world—such as who might own the copies of ourselves. *Orphan Black* handles these philosophical tensions beautifully, even delving into some of the politics of patenting human genes, while, at the same time, not settling for easy answers.

PART II

“If we’re genetically identical, do you get that little patch of dry skin between your eyebrows?”

5
The Human Being in the Age of Mechanical Reproduction

DANIEL MALLOY

> The presence of the original is the prerequisite to the concept of authenticity.
>
> —WALTER BENJAMIN, "The Work of Art in the Age of Mechanical Reproduction"

We live in an era of technological and biotechnological breakthroughs. One of them is explored by *Orphan Black*—the mechanical reproduction of the human being. Many are frightened by the prospect, as people always are in the face of something radically new. But rather than be frightened by the possibilities presented by cloning and other biotechnologies, we need to embrace them.

Human nature has always been flexible. Dr. Leekie, flawed though he was, was right to say when we first met him in "Variations Under Domestication" that present and near-future biotechnological advancements give us "the opportunity at a self-directed evolution . . . that's not only a choice but a human right." Technologies like cloning don't change that. All they do is increase the range of possibilities and enhance our consciousness of that flexibility. They give us a greater ability to decide what we are—and hence, a greater responsibility for what we are. Cloning and other biotechnologies force us to take greater responsibility for ourselves, which ultimately only serves to make us more human.

A similar sort of transformation has happened in the world of art over the past century and a half. Living in the age we do, we forget that not only is the mechanical reproduction of human beings a recent development—mechanical reproduction of anything at all is still rather novel. We're still coming to terms with the ways that mechanical reproduction changes what is reproduced. One of the best guides in grappling with this puzzle is philosopher and literary critic Walter Benjamin (1892–1940). In "The Work of Art in the Age of Mechanical Reproduction," Benjamin speculated that technologies like photography, film, and phonographs, which allowed for the mechanical reproduction of works of art, would change art in two fundamental ways. First, it would strip works of art of what he called their "aura"—the unique connection each distinct work maintains to its origin. Second, it would lead to a "democratization" of art, by making what had formerly been the province of a few widely available to all—and hence encouraging more people to engage in and make art.

Benjamin's thoughts on the mechanical reproduction of art can serve as a guide to the mechanical reproduction of the human being. Cloning and other biotechnologies have the potential to strip us of our "aura," but also to "democratize" human nature itself. Tony Sawicki, the transgender clone, demonstrates the barest inkling of the possibilities for altering and redefining ourselves and, by extension, what it means to be human.

The Genuine Article

Until around the middle of the nineteenth century, works of art were relatively simple things. It's never been easy to define what makes them works of *art* exactly, but until the invention of photography, and the subsequent developments of sound recording and motion pictures, works of art were at least easily identifiable as *things*. A sculpture, a painting, a story, a song, or even a performance were all particular things created in a specific time and place. So long as they existed, they remained connected to that time and that place. One of Felix's paintings is a singular, specific object in the world that has a location and a history that is unique to it. Whatever happens to it, wherever it's moved and whomever possesses it, the painting will always retain a connection to the time and place and circumstances of its creation. This particular connection is what Benjamin referred to as the "aura" of the work of art.

The aura of a work of art is something that we do not often give a great deal of thought. The central and important aspect of a work of art is the immediate experience of it, something which is not necessarily connected to its past or origin. A person exposed to one of Felix's paintings or one of Kira's drawings for the first time may be moved even if they have no other information about it—that is, even without knowing that what they're experiencing is this particular work produced by this particular person in this time and place.

But all art has context. This context is part of what makes art more than just a hobby that humans engage in occasionally, and turns it into an institution—a central aspect of human life, akin to politics or religion. It is in thinking of art as an institution that the aura of the work of art becomes essential. The aura of a work of art is what grants that work its authenticity. The aura is what distinguishes the genuine article from a mere copy or imitation. There's a kind of indefinable difference between a Felix Dawson original and a forgery or knockoff, no matter how good.

If art was not an institution, if it was just about the immediate aesthetic experience of individual works, then there would be no difference—or at least no important difference—between Felix's painting and a print based on it, or a forgery for that matter. It is the original's aura that gives it its special status—the fact that it was created by the hands of the artist himself makes it unique and exceptional, even though the aesthetic experience of it is no different from the experience produced by a good reproduction of it.

The Authentic Human

Human beings are not works of art. There are many differences, of course, but one of the central ones—and certainly a central one for *Orphan Black*—is that humans are not the sorts of things that can be owned. Things can be owned; people can't. When Cosima reveals that "We're property. Our bodies, our biology, everything we are, everything we become, belongs to them" ("Endless Forms Most Beautiful") she is, in essence, revealing the ways that she and the other clones have been stripped of their personhood. They've been treated as mere things.

What sort of parallel is there between works of art and human beings—if works of art have auras, do we as well? Not exactly, because so far human beings are not *made* things in the way that works of art are. But we are beings who begin to exist in particular times and places under particular conditions. We are all born to particular parents as parts of particular families, and grow and mature under specific sets of circumstances. Even the members of the clone club are still born in specific times and places to specific (surrogate) mothers.

Philosopher Hannah Arendt (1906–1975), a contemporary and friend of Benjamin's, coined the term "natality" to describe this fact about us. Our natality is, like an artwork's aura, our connection to our unique beginning.

It also indicates how, regardless of whatever else happens in the course of our lives, our beginning shapes and defines us in innumerable ways. Just as the work of art's aura makes the original "authentic," our natality makes it possible for each of us to be uniquely ourselves—to be authentic or genuine. Only by embracing our pasts, including our origins, do we become ourselves. So, for example, it was important for Sarah and Helena to meet their surrogate mother, Amelia ("Unconscious Selection" and "Endless Forms Most Beautiful"). Their shared birth mother, as part of their natality, creates a link between them that is difficult to sever.

There's a significant difference between the artwork's aura and the human being's natality, one that comes down to an important distinction between human beings and mere things. The artwork has no choice in the matter. Felix's painting cannot declare, as Sarah does in the show's teaser, "I am not your property." Nor, for that matter, can it consent to being property, as Alison nearly does in "Endless Forms Most Beautiful."

A work of art is completely defined by its aura; the human being, on the other hand, can do a number of things regarding her own natality. She can embrace it or run from it; accept it or deny it; acknowledge it as a part of her or allow it to completely define her. The choice is up to the individual person. For example, consider Sarah's and Alison's distinct reactions to their origins. For Sarah, it's a concern because of the problems it may present in the future. For Alison, on the other hand, it's a matter of shame—she's embarrassed that she's a clone.

Regardless of how each of us responds to her own natality, however, just as with the work of art's aura, it's a fact about us. We can choose how to deal with that fact, but we cannot choose not to deal with it. It's a given; an unchangeable, immutable thing that confronts each of us in our attempts to go about our lives.

Prints, Recordings, and Clones

Works of art initially began to lose their auras in the mid-nineteenth century, with the advent of photography. In rapid succession, sound recording and movies further eroded the link between works of art and their origins. The digital revolution has further eroded the aura to the point where it is little more than a fossil. Each of these technologies has undermined the aura of the work of art by calling into question the importance—indeed the very idea—of authenticity.

Take the photograph, for instance: when a photographer takes a picture, she has not created a work of art—not yet. Later the film from the camera must be developed to get a photograph. Now, the question raised by the photograph is where is the work of art? It isn't the individual print of a photography—that can be reproduced as many times as we want. And it's not the film negative—that isn't what the consumers or audience of the photograph take in.

Or consider Dr. Duncan's copy of H.G. Wells's *The Island of Dr. Moreau*. It's a copy, a print. The physical object has no particular connection to Wells himself. The work of art involved, whatever that is, is just as easily accessible in a variety of forms, including numerous copies that are nearly identical to Duncan's (without his added notes, of course). This endless reproducibility, which has always been a component of linguistic artworks like stories and poems, deprives the original—again, whatever that is—of its significance.

There is no "authentic" photograph or "genuine" *Island of Dr. Moreau*. In the photograph's case, even its connection to what it portrays is tenuous at best. The same holds for sound recordings and motion pictures. And the digital revolution has done even more to undermine the notion of the authentic work of art. Authenticity and aura go out the window with the entrance of mechanical reproduction because mechanical reproduction rips works of art free from the context of their creation.

But what about works that are not photographs, sound recordings, or movies? Don't they retain their auras? To an extent and for a while, yes. But it becomes less and less central to the institution of art. Once photography and motion pictures are acknowledged as artworks on a par with paintings and sculptures, the focus of a work's value shifts away from its authenticity to its immediate impact. Reproduction largely frees works of art, including non-reproducible works, from their dependence on context. As Benjamin explains it, the exhibition value of the work of art overtakes its cult value. This leads to a revolution in what a work of art is.

Leda, Castor, and Authenticity

Are clones human? This question is the human equivalent to the question of whether mechanically reproducible works—photographs and movies—are works of art. And the answer has ramifications that are just as radical for the question of what it means to be human. Determining whether Sarah and her sisters are human—or merely, for example, property—is the core conflict of *Orphan Black*, a question that the show answers with an emphatic yes.

But if Sarah and the rest are human, that means that the origin of the individual human being takes on diminished importance. Cloning is merely the next step in this process. Natality began eroding with the advent of fertility treatments, particularly with procedures like in-vitro fertilization and surrogate motherhood. Humanity is already becoming unmoored from its origins, from its natality. Sarah the clone is just as human as Felix the foster son or Kira the natural-born daughter.

But, it may be objected, human beings are still born in particular times and places and circumstances. Even the clones in *Orphan Black*, while their conceptions may have taken place in circumstances that were far from conventional, were born in the old-fashioned way, or something closely approximating it. Sarah and Helena aren't just fellow members of the Clone Club; they're twins because they shared the same birth mother, Amelia. All of the clones had surrogate mothers who bore them in somewhat natural pregnancies and birthed them in something like the usual way.

This fact represents a remnant of natality, not the full thing. It explains some of the slight biological differences between the clones—such as Sarah's and Helena's fertility and Helena's status as a mirror twin, for instance. But the apparently natural birth of the clones is no longer a necessary part of their being. It simply represents a present limitation in the technology available for cloning.

Even as I type this, scientists are working on developing artificial wombs, which would be capable of carrying a fertilized egg from conception to gestation without the use of even a surrogate mother. Should such technology become a reality—and there's little reason to think that it won't—it would be the death knell for natality. The origins of clones—

and by extension all other human beings—would become irrelevant. The sorts of horrific baby-farming witnessed at the Prolethean farm in "Governed as It Were by Chance" would be unnecessary.

Perhaps, then, we could rescue natality and human nature by denying that clones are human. In their own ways, the Proletheans as well as the powers behind Project Leda and Project Castor choose this path. In each case, the clones are viewed as things: as tools, as experiments, and as weapons; but never as people. The reasoning in each case seems to be that clones are products, manufactured beings that were made to serve a specific purpose. Thus their origins determine their essence, and rob them of their autonomy. If our natality is essential to what makes us human, and Clone Club members lack it, then Sarah and her sisters are not human. Clones are property. However, this would be playing a game that's already lost. The instant cloning—particularly on a mass scale—becomes a reality, natality is done for as a defining aspect of human nature. Once the question is raised it is automatically answered.

To see why, imagine the ramifications of denying the humanity of clones. First, it would become necessary to have some easy way of identifying who was and was not a clone. Remember, Sarah had no clue she herself was a clone until she encountered Beth on that train platform. This leads to an emphasis not only on natality, but the ability to prove your natality. Your whole identity—whether a clone or not—becomes identified with and defined by your origin. In turn, non-clones lose their ability to define themselves just as much as clones do. The Proletheans, for example, by denigrating the clones, are led to pride themselves on their status as natural-born—and little else. To deny clones their humanity undermines an essential aspect of all humanity. The freedom of the naturally born and the artificially conceived stands or falls together.

Just as has happened with art, in the name of redeeming anything human, we must embrace the humanity of clones as much as the natural born. The alternative is to renounce the idea of humanity all together. Either Sarah and her sisters are human, or no one is. But if the members of the clone club are human, then humanity and natality can be separated. Thus, as happened with auras and artworks, the understanding of what it means to be human must shift to account for this new reality.

Democratizing Art

Art's loss of aura strips it, in large part, of its cult value, of its connection to and dependence on ritual. What remains without those? Just exhibition value—the value the work of art has in being experienced firsthand. Art has always had exhibition value, but mechanical reproducibility makes exhibition value its only value. Movies and photographs are not objects to be owned in the same way that paintings and sculptures are. They must be exhibited, displayed, experienced. Without some public display, they lose whatever value they have.

What makes Kira's drawings and colorings valuable? We can admit that Kira is a lovely child while also acknowledging that she is not Rembrandt. The value in Kira's artworks has little to do with their aesthetic qualities, and everything to do with their origin. People who prize Kira's art do so because it is Kira's, not because it's good. The little girl's artwork has cult value because of its aura, but it generally lacks exhibition value.

Imagine that Felix created a painting, his masterpiece, which he then gave to Sarah on condition that she never show it to anyone else. The painting is thus robbed of its exhibition value, but in turn, its cult value skyrockets. No one besides Felix and Sarah knows what it looks like, so it has no exhibition value. But, the fact that it is a secret, Felix Dawson original, only enhances its cult value.

We can't imagine the same happening with a mechanically reproducible work of art. Imagine a similar scenario: suppose everyone involved in *Orphan Black* got together and produced a secret episode. Following the pattern established by Jerry Lewis and his work *The Day the Clown Cried*, they donate the episode, unaired and unscreened, to the Library of Congress on condition that it is never shown. Ever. To anyone. Under those conditions, the episode wouldn't be an episode, or a work of art at all. It would be just a bizarre thing in the Library of Congress. Robbed of their exhibition value, mechanically reproducible works of art cease to be art.

The reduction of aura, and in turn, of cult value, however, is not something to be lamented. As Benjamin understands it, by making works of art necessarily public things in a certain sense, mechanical reproduction leads to a democratization of art. No longer the province of the wealthy and the bored, art becomes accessible to all—both in terms of enjoyment and in terms of production. Anyone with a smartphone today has better camera technology at their fingertips than even existed in the first few decades of photography. Mechanical reproduction liberates art from the strictures of institutions.

Not Our Usual Identity Crisis

We human beings are similarly liberated by being stripped of our natality. Birth is no longer destiny. Biology now presents a range of options and possibilities instead of a sentence. The members of the clone club demonstrate this truth as no one else can. Given nearly identical biological starting points, they have become vastly different and distinct people—more than that, they have chosen to become who they are.

One of the often voiced concerns about biotechnology is that it has the potential to rob future generations of their freedom. Some argue that a child genetically engineered to be a great athlete or scientist has no choice in the matter. Rachel Duncan, born a clone and raised a clone, is "proclone," as Felix calls her.

However, this concern demonstrates a basic misunderstanding of both the biology and the technology involved. Biologically, there are limits to what can be determined about us even with the most sophisticated techniques—as Sarah's and Helena's fertility demonstrates. They, like all the other clones, were designed to be infertile. And yet, through some quirk they both wound up able to have children. A child may be designed, for instance, to be an excellent basketball player. His genes would then be selected to make him tall, strong, fast, and well-co-ordinated. But whether he will enjoy basketball, or even be good at it, cannot be predetermined. At most genetic engineering gives the child a particular set of tools—perhaps a better set than the genetic lottery of natural conception would have, but still just tools. The best tools do not, however, make the best artist or craftsman.

The second issue with this objection is that it fails to acknowledge that the engineered person will also have access to the same technology that engineered them. Here the case of Tony Sawicki is exemplary. Tony the clone was born Antoinette. Antoinette was not comfortable being Antoinette. Taking advantage of currently available biotechnology, Antoinette chose to become Tony. Having gone through his transformation, Tony is perhaps the best equipped of the clones to deal with the discovery that he is, in fact, a clone. After a brief shock on meeting Sarah, Tony responds, "I did all that work a long time ago. There's only one Tony and you're not me, sucker."

Tony is Tony, but more than that, Tony chose to be Tony. So long as natality rules the day, we can only choose our identities by embracing what we are born to. But with the end of natality, we also have the end of pre-given identities. The death of natality is also the death of destiny—and the birth, for the first time, of true autonomy. Biology doesn't determine our destiny. Thanks to biotechnology, biology doesn't even determine biology.

While being stripped of their auras likewise stripped works of art of their authenticity, being stripped of our natality only enhances human authenticity. For the first time ever, we have the opportunity to truly be ourselves—and to decide for ourselves who that is. Sarah is never surer of who she is, or more committed to being and remaining herself, than when dealing with her sisters. Being a member of the clone club requires being a clone, but also being *you*.

The core of the human being, stripped of natality, is autonomy—the ability to make choices for ourselves. The term autonomy comes from two Greek terms: *autos*, meaning self, and *nomos*, meaning law. An autonomous being is thus one that can give itself laws—or, more simply, rule itself. Biotechnology only expands our ability to rule ourselves. We are no longer subject to the random dictates of inherited genetic sequences and hormones and ancestral lineage. Nearly the same set of genes can create a Sarah, a Helena, an Alison, a Cosima, a Rachel, a Beth, and a Tony. Biotechnologies of various sorts, once they become widely available, represent the ultimate form of democracy. The people can finally rule themselves—and even their own bodies—in ways we've only ever dreamed of.

Embracing the End

The biotechnological revolution presented to us by *Orphan Black* can seem frightening. It is frightening. Freedom always is. But more than that, we're just on the verge of this revolution. It's a delicate time, when a wrong choice can have disastrous consequences. There are two wrong choices confronting us.

The first is the choice advocated by biotechnological Luddites—those who would like us to stop pursuing biotechnologies in the name of preserving the human status quo. The problem that the Luddites don't seem to see is that the biotech genie is already out of the bottle. Natality died in 1978 with the birth of Louise Brown—history's first "test tube baby."

The second choice is the one advocated by the officials of Projects Leda and Castor—the choice of treating the bioengineered as something less than human, as property. To take this path would ultimately degrade all of humanity. Once one human is declared property, it's only a matter of time before all humans are property. Just look at the attitude Rachel demonstrates toward others. Her coldness and distance, her manipulativeness and selfishness, are not the result of being a clone, but of thinking herself as a clone, a mere thing, rather than a person. If she's just a thing, then so is everyone else.

So, all things considered, our best option is to embrace the biotechnological revolution and celebrate the end of our natality. Don't mourn the aura. We're better off without it. We're not less human, but more human, once we take conscious control over our own nature. And Dr. Leekie was right in "Variations Under Domestication": it's "not only a choice, but a human right." The biotechnological revolution gives us the right to be human, and the right to decide exactly what that is.

6
One Clone from Another

ERIK BALDWIN

Orphan Black depicts the Leda clones coming to terms with their clone status. The clones, and we as the audience, are struck by their many similarities and differences.

Beth, Sarah, Helena, Alison, Cosima, Rachel, Tony, Katja, and Krystal are each unique characters, each has distinguishing characteristics or features that clearly distinguish them one from another. As the clones become better acquainted and build significant relationships with one another, each seems to take on some of the characteristics of the others. They seem to have an innate ability to impersonate one another.

For much of Season One, Sarah's impersonation of Beth is good enough to convince Beth's detective partner, Arthur, that she *was* Beth. Sarah impersonates Alison in order to handle her house party ("Variations under Domestication") and family day in rehab ("Knowledge of Causes, and Secret Motion of Things"). Rachel's impersonation of Sarah took even Felix by surprise ("Things Which Have Never Yet Been Done"). To trick Dyad, Alison impersonates Sarah while Sarah impersonates Rachel ("The Weight of This Combination") and Ferdinand is none the wiser. Cosima's and Helena's impersonations are rough but get the job done. Helena evades detection by impersonating Sarah's impersonation of Beth, managing to trick "Beth's" not-so-secret admirer Raj ("Effects of External Conditions").

With Alison busy delivering money to Pouchy's niece, Felix forces Cosima to stand in for Alison to do photo ops and to deliver her speech. While Pouchy and his crew seem somewhat suspicious, that's probably just their default setting. They seem to accept Helena as Alison in "Community of Dreadful Fear and Hate", at least until they threaten Alison and Donnie's children. Apparently, Alison's training as an actress and Sarah's experience as a con-artist prepared them for their roles much better than Helena's time at assassin school and Cosima's courses in making crazy science.

The Metaphysics of Persons

How are we to understand and account for the relations of similarity-and-difference that the clones bear toward one another? What distinguishing features make them unique persons? In asking such questions, we are inquiring about which features or characteristics individuate each particular clone as a distinct being or entity. Our main question, then, is "What metaphysically individuates one clone from another?"

One answer to this question is that each clone is a personal substance. Following John Locke (1632–1704), to be a person is to be "a thinking, intelligent being, that has reason and reflection, and can consider itself as itself, the same thinking thing in different times and places" (*Essay concerning Human Understanding*, II, xxvii, 9).

According to a traditional understanding of the notion, the roots of which date back to the ancient Greek philosopher Aristotle (384–322 B.C.E.), persons are metaphysical substances that have properties. For example, consider the sentence "Sarah is a Leda clone." Sarah is not just the grammatical subject of that sentence, but the metaphysical bearer of a feature or characteristic that some things in the world have and most do not, namely, the property of being a Leda clone.

According to this approach, substances have properties and while properties have real existence, they cannot exist apart from or independent of the substances in which they inhere. Moreover, substances endure; they exist over time throughout processes of change. For instance, 'being five feet four inches tall' is a property that is predicated of each of the Leda clones. When they were children, however, they were much shorter.

As substances, although each clone loses an old and acquires a new height property as she matures, each remains the same individual over the course of those changes. Pulling all this together, on the traditional view, to be a person is to be a unified, substantial whole that is self-identical throughout processes of change.

One influential account, associated with Plato (427–347 B.C.E.) and Descartes (1596–1642), is that human persons have a non-repeatable nature, essence, or property. For instance, 'being Rachel' or 'being Alison' are non-repeatable, non-sharable properties uniquely had by and essential to their bearers. Such properties are called individual essences. Rachel cannot exist without having her respective individual essence and nothing other Rachel could possibly have that individual essence. Naturally, each of the clones has an individual essence that distinguishes her from each of the others.

Some philosophers, including David Hume (1711–1776), object that things like individual essences are dubious. How could we know that there are such things? How could we possibly be acquainted with them? Such properties can seem mysterious or even magical.

Idealists, such as Bishop George Berkeley (1685–1753) and Georg Hegel (1770–1831), maintain that humans are mental substances, that human persons are identical to an immaterial soul or mind. One problem with soul or spirit being that which individuates persons is that it's difficult to clearly demarcate distinct souls. How do we attribute mental and perceptual states to one soul and not to another? And on what grounds can we tell whether a unique soul is attached to each human person or whether there is one universal 'world soul' that expresses different aspects of itself in different people?

Another proposal is that being composed of different bits of matter individuates human persons. According to one variation of the view, human persons share in common many of the same universal properties, such as 'being a rational animal', 'being a living thing', 'having skin and bones,' and the like, and what separates one individual from another is the compresence (or the bringing and tying together of features in a substance) of various universal properties in a particular material body that persists over time as a self-identical, unified subject of predication. This view has its roots in Aristotle, was accepted by St. Thomas Aquinas (1225–1274) and many other medieval philosophers, and has many contemporary advocates.

One problem with this view is that, so scientists say, humans switch out their matter with other bits of matter as they age and mature. But how can a human person persist if all of their material bits are replaced by new ones over time? To address this problem, Locke proposes that psychological continuity and memory ground personal identity. But this answer generates other problems. For one, it implies that when a person is unconscious, such as when asleep or in a coma, she ceases to exist but is somehow reconstituted upon awakening. But something that has a discontinuous or choppy existence isn't something that persists over time! Other problems involve cases of consciousness swapping and reduplication. Suppose that Dyad or Neolution can make multiple copies or backups of a particular clone's consciousness and memories and can download those copies into other bodies. That this is conceivable goes to show that having the same memories or experiences isn't enough to ground personal identity.

Another problem with these proposals is that there are reasons to doubt that humans are substances. Descartes defines a substance as "a thing which exists in such a way as to depend on no other thing for its existence" (*Principles of Philosophy*, I, 51). He distinguished created and uncreated substances. God is the only uncreated substance, for only God depends on nothing else in order to exist. Humans are created substances in that their existence depends on nothing other than God. Benedict Spinoza (1632–1677), a contemporary and early critic of Descartes's philosophy, argues in his *Ethics* and *Principles of Cartesian Philosophy* that if human persons are dependent on God for their existence then, strictly speaking, they aren't substances after all.

Hume questions the meaningfulness of the idea of substance. He maintained that all ideas arise from experience. He argues in *A Treatise of Human Nature* that if we can't trace back an idea to an original sense impression, then we don't have good reason to think that it is meaningful. And since we can't trace the idea of substance back to an original sense impression, we shouldn't believe that there are any such things. Hume goes on to argue that anything that is clearly conceived as a distinct and separate thing may possibly exist as such.

For instance, we can conceive the very same visceral, raw feeling of fear or anxiety as being had by any one of the clones. And we can conceive of particular properties or features of clones as existing separate from a particular clone. We can conceive Helena's hairs on Cosima's head and vice versa. We can conceive the creepy voice of Helena's imaginary 'pet' scorpion Pupok even though it doesn't have a mouth suitable for forming words and scorpions don't have creepy voices. Apparently, Helena ate a real scorpion in "Certain Agony on the Battlefield" she took to be Pupok. But that doesn't necessarily mean the end of Pupok! Presumably, it's not that easy to kill an imaginary being. Pupok might reappear next season as though nothing has happened. On the basis of such considerations, Hume concludes that we don't need to assume the existence of substances as the sorts of things in which we find perceptions, ideas, or properties.

The Psychology of Personalities

When we consider our question from a psychological perspective, other problems arise. Why did Alison break up with Jason? Why did she marry Donnie? Why did Beth become a detective? Why did Sarah take up a life of crime and get involved with Vic? Why does Cosima love making crazy science so much? Why did Helena dye her hair blonde? Answers to these questions need to make reference to differences in their respective upbringings, experiences, influences, and choices. We may readily imagine that the members of Clone Club would have much different personalities if these factors were significantly modified.

Psychological identity and metaphysical identity aren't the same but they are related. Can human personalities radically change without the loss of one's metaphysical identity? Let's imagine Alison and Cosima were to somehow completely trade personalities. Could each of them remain the same self-identical unified whole despite undergoing that change? Perhaps. On the other hand, if Dyad were to take Helena's personality—her thoughts, memories, experiences, and all— and try to 'download' them into a scorpion, we can be confident that Helena wouldn't survive *that* change.

A scorpion's brain just isn't a suitable receptacle for human consciousness. The most we can reasonably expect is that certain traces of Helena's experiences or memories might take hold in the scorpion's brain. In contrast, Felix might make a few small changes here and there to one of his paintings without its being destroyed or turned into another piece entirely. But should he make too many fundamental changes, we should probably say that the old painting is destroyed and a new painting has come to be. Humans are more complex than paintings, but we can draw similar conclusions. If the personality traits, memories, mental states, and abilities associated with the Leda clones are metaphysically essential properties, then a given clone wouldn't survive the loss of those properties.

The Buddhist Angle

So far, we've been framing and answering our question within a traditional Western philosophical framework. For another perspective, consider how Buddhist philosophers address our question. We start with the Four Noble Truths, which capture the core elements of the Buddhist path. These are the teachings that Vic is referring to when he tells Alison that he has reformed and is "walking the path of the Buddha" ("To Hound Nature in Her Wanderings").

The first Noble Truth is that everything is *dukkha,* or suffering. The second is that suffering is caused or conditioned by desire or thirst. Third is that there is a path to the cessation of suffering. The cessation of suffering is *nirvana*, which isn't a place or thing (or a popular 1990s grunge band from Seattle), but a state about which Buddhists are reluctant to say anything positive. The Buddha says that our existence is analogous to that of the flame of a burning candle and *nirvana* is analogous to the blowing out of that flame. The fourth Noble Truth states that the path to *nirvana* is The Middle Way between self-indulgence and asceticism, also called the Eightfold Path, described in the chart below:

1. **Right views or understanding: basically, accepting the Four Truths.**
2. **Right thought or resolve: developing attitudes of freedom from desire, friendliness, and compassion; abandoning hatred, sensual desire, and abstaining from causing injury.**
3. **Right speech: not telling lies, avoiding divisive speech, harsh speech, including hateful or abusive language, and frivolous talk, such idle chatter or gossip.**
4. **Right action: avoiding killing, stealing, and inappropriate sexual conduct.**
5. **Right livelihood: not engaging in an occupation that causes harm to humans or animals.**
6. **Right effort: mental cultivation, transforming one's mind by replacing negative thoughts with positive and wholesome ones.**

7. **Right mindfulness: developing constant awareness of one's body, feelings, moods and mental states, and one's thoughts; the elimination of hindrances, such as sensual desire, sloth, worry, and anxiety.**
8. **Right concentration: developing mental clarity and calm by concentrating the mind through meditational exercises.** (Adapted from Prebish and Keown, *Introducing Buddhism*, pp. 52–54)

Since our focus is on metaphysics and not ethics, for our purposes, we aren't concerned with the third and fourth Noble Truths. We consider aspects of the first two Noble Truths in greater detail below.

Dukkha includes ordinary suffering, metaphysical impermanence, and conditioned states that are *svabhāva*, or empty of own-being. Buddhists maintain that nothing persists or endures over time. Accordingly, they don't think that there are any substances. One application of this view is the doctrine of *anātman*, which states that there is no such thing as an eternal or unchanging self. What we call persons are 'reduced' to conditioned states, or *skandas*, that have momentary existence, of which there are five: body (*rūpa*), feelings/ sensation (*vedanā*), sense-based perception/conception of objects (*saṃjñā*, mental formations, dispositions, or volition (*saṃskāra*), and consciousness awareness (*vijñaña*).

While Buddhists don't think that persons are metaphysically basic, for the sake of convenience and easy communication, they think it's okay to talk as though they are, as long as we realize that such statements don't depict the way things really are ultimately. For example, we talk as though cars are metaphysically basic. But, if we think about it, we recognize that to be a car is for certain parts, namely wheels, doors, seats, engine components, and the like, to be arranged in a car-like way. We don't really think there is something over and above the parts that are so arranged. We don't think that a car is a substance. (Early Buddhists used more appropriate examples such as chariots.)

Alison and Donnie, having bought out Ramon's drug supply ("Transitory Sacrifices of Crisis") in order to pay the bills and to fund Alison's campaign for school trustee, set up a front business selling homemade soaps in nicely designed wooden crates. While Alison and her customers may talk as though each package is a thing, everyone knows they're really just collections or bundles of various more basic things, including wood, decorative paper, soaps, and various illegal drugs labeled sertraline (Zoloft) and purple drank (promethazine plus codeine). Similarily, Buddhist philosophers maintain that human persons are collections of more basic things, too.

Co-dependent Origination and Contradictory Self-Identity

Buddhists maintain that every thing is dependent on external causes and conditions. This view is called the doctrine of *pratītya-samutpāda*, or dependent origination. According to this view, everything is what it is on account of being related to everything else. Nothing has an intrinsic nature or essence. It follows that none of the clones is a metaphysically distinct person independent of her relationships with the others.

Japanese philosopher Nishida Kitaro (1870–1945), elaborating on the logical ramifications of dependent origination, formulates a principle of contradictory self-identity according to which a thing is itself only by being other than itself. According to Nishida, an individual thing, call it *A*, and another individual thing, call it not-*A*, overlap such that *A* is what it is only insofar as it is related to and expresses something about not-*A*. Nishida refers to this as the logic of *soku*. Stated as a formal rule of inference, we have "*A* is *A*, and yet *A* is not-*A*; therefore *A* is *A*." This principle doesn't imply anything nonsensical or logically contradictory, such as that Cosima both has and doesn't have dreads. Rather, Nishida is saying that in order for a thing to be what it is, it must contain elements of what it is not or it wouldn't really be what it is.

Each Leda clone is going through the biological process of living and each clone is going through the biological process of dying. And although the properties "is living" and "is dying" are conceptually opposite, living creatures just are dying creatures—the concrete biological process called living is no different than the concrete biological process called dying. Put succinctly, to live is to die. That is, you can't really be a living thing unless you're also a dying thing—to live is to live and yet to live is to not-live; therefore, to live is *to* live."

Another Japanese philosopher, Seiichi Yagi (born in 1932), provides a complimentary analysis of dependent origination. What he has to say helps us to understand Nishida's views about contradictory self-identity. Adapting one of his examples, consider the prison cells occupied by Sarah and Helena. A wall divides the cells from one another. The surfaces of the wall are constitutive parts of two different prison cells. You wouldn't have two prison cells without the wall that separates them. And you can't have a prison wall unless you put materials up such that you divide a space between them to make two cells. In this way, prison cells and walls condition one-another—each is what it is but only on account of being related to what each is not.

Seiichi talks about a house with a lush garden. The plants in the garden are part of nature but also a living area of a family's home. The home stands in contrast to nature. The garden is a part of nature that has been domesticated. But to completely eliminate nature from the garden would be to eliminate the garden. So a garden is a part of nature that has been made into a living area for a family yet also an area that also remains part of nature. These and other examples illustrate Nishida's view that one individual thing and another individual thing overlap such that each is what it is only insofar as it is related to and expresses something of what it is not.

We're always going through processes of change and always contain characteristics or features of other individuals in order to be the individuals that we are. Thus, if we're to speak about ourselves accurately, we must refer to ourselves as what we are *and* as what we are not simultaneously and in precisely the same respects.

The Clone Dance Party

Let's consider how Nishida's principle of contradictory self-identity can help answer our question using the clone dance party scene in "By Means Which Have Never Yet Been Tried" as a focal point.

Recall the scene. Sarah takes Helena to Felix's flat to meet Alison and Cosima for the first time. Felix introduces Helena to Cosima. She drops the needle on the electronic reggae song "Water Prayer (Matt the Alien Remix)" by Adham Shaikh and starts dancing with Felix. As Felix motions her to the impromptu dance floor, Cosima moves her arms about, bobbing and swaying. Sarah is next to get up and groove. Her dance is more determined and aggressive yet uses many of the same core moves as Cosima. Alison and Helena sit nervously on the couch.

Felix pulls Alison up from the couch. Reluctantly and with reservation, she makes some modest moves. At the same time, Helena decides to get up and dance. While neither seems to know just what do to with themselves at first, Helena soon lets loose and dances about wildly, flailing her arms and legs. Alison slowly gets more into it but continues to dance in a self-conscious, almost cautious way.

As the dance party continues, each clone's movements are subtly influenced by all of the others. At one moment Alison mirrors Cosima's trippy hand motions. At another Sarah jumps up and down like Helena. Each clone's dance is their own but each clone's dance is what it is on account of incorporating and mirroring certain elements of the moves of the other dancers. Applying Nishida's logic of contradictory self-identity to the dance party scene, Alison's moves are her moves and yet they are not her moves, but Helena's, so they are truly Alison's moves. Similarly, Cosima's moves are her moves and yet they are not her moves, but are Sarah's, so they are truly Cosima's moves, and so on.

This analysis may be extended to many other ways in which the clones are interrelated. For instance, later on in the same episode, Cosima and Sarah lay side-by-side holding hands on Felix's bed. Discussing their clone status, Sarah notices Cosima's nautilus tattoo and remarks, "we're so different, all of us." But even as they talk about their differences, the similarities between Sarah and Cosima are evident. Numerous other scenes display this dynamic of similarity and difference.

Our analysis of the clone dance party provides another answer to our question, "What metaphysically individuates one Leda clone from another?" Recall the inference rule, "*A* is *A*, and yet *A* is not *A*; therefore, *A* is *A*." Substituting the variable *A* with the name of an individual clone we have the sentence, "Alison is Alison yet Alison is not-Alison; therefore Alison is Alison." Paraphrasing and amplifying the meaning of the sentence, Alison is the unique individual she is on account of being extrinsically related to things that are not-Alison, specifically, her clone sisters. That she stands in these relationships with her sisters is what makes each clone the unique individual she is. Each clone has a unique personal identity, but that identity is not an intrinsic feature of her being. Nishida's principle of contradictory self-identity offers a way to understand how the members of Clone Club could be metaphysically individuated.

7
I Am and Am Not You

JEREMY HEUSLEIN

Sarah is back from being away; she has argued with someone on the phone about Kira, her daughter, from whom she is being kept away. She seems to be on the run and is waiting around a train platform for some reason, presumably to catch a train, or maybe to start running away again. After a moment, she catches sight of a young woman, about her age and about her size.

This woman's sobs are audible. She seems in distress, and for Sarah, there is something about her, and she is drawn closer to her. As she approaches, with the woman's back still to her, the woman sets down her purse and takes off her shoes and her coat. Then she turns around. In an instant, Sarah's world is thrown open (beginning the whole course of its being thrown open more fully throughout the show). She looks in shock (horror? fear? disorientation?) at *her* face on this woman looking right back at her.

The woman's expression does not change, not really. It's as if she is unfazed by herself looking at her. In fact, it seems that seeing her own face on this young woman, dressed in almost an opposite fashion, encourages her to keep walking towards the tracks and into the oncoming train. She does, and so she kills herself. Sarah sees the woman to become known to her as Beth—but in this moment completely unknown to her besides sharing a face—

step in front of train and die. Sarah is thrown from shock to shock.

I have tried to imagine, as I assume most fans of the show have as well, that moment in my own perception. It's difficult, particularly because (as far as I know) I'm the only me there is. I'm the only one with this face and more emphatically this body. This enables me to understand to some degree the disorientation Sarah must have faced. The encounter between Sarah and Beth contains two elements that deserve attention: face and body. These women share a face, but only one dies, only one body is killed (and we do not see her face again, except on videos of her past). Sarah then is launched into a world of intrigue, of scientific experimentation, of death, and of playing god that is *Orphan Black*. From this moment, all others follow. Her identity begins to unravel and respin itself. This happens in a very bodily way.

In philosophy, there's a school of thought called phenomenology, which emphasizes a rigorous investigation of experience to understand philosophical problems. Edmund Husserl, the founding father of this school, gave it the mantra: "To the things themselves!" Husserl also investigated, as a part of going back to the things themselves, the body, as it is lived and experienced.

For him, there were two words to describe the body, one meaning the body in its bare materiality or physicality (the body as a thing) and the other expressing the sense of livedness, the experiencing body, often translated from the German as the "lived-body." In Husserl's understanding, every "I" inhabits its lived-body primarily, with its "bare materiality" body being an abstraction (the scientific disciplines treat the body largely in this way). In fact, Husserl went as far as saying, in an analysis on perception, that the lived-body is the absolute "here," in relation to which all "theres" exist. In other words, where I am my lived-body is, and all distance and space and objects refer back to this positioning. This is where we first begin to understand identity.

Sarah's dramatic thrust into the world of clones, the other "hers," begins with a lived-encounter. Encountering another body like hers and then another and then another unravels Sarah's sense of identity. In phenomenology there are three approaches:

1. the "static," in which an object is described in its essential identity or characteristics, for example, a pencil is described in "static phenomenology" when the focus is on its spatial or substantive characteristics—color, weight, material construction, and other basic elements such as "thinghood" rather than animality;
2. the "genetic," in which an object is described more in terms of its temporal constitution, its identity throughout history, for example, the pencil's own history, when it was made, whether it is sharpened, the process of sharpening, and so forth; and
3. the "generative," in which the object is seen in light of the history of the world and contributes to that unfolding history, for example, the pencil seen in light of the history of writing, of literature, the creation and the destruction of this particular pencil, and the words written by it.

In this way, in a "generative phenomenological" way, an object or a person cannot be reduced into its own history. When Sarah meets Beth's eyes, her identity unravels in its generative stream. Statically, she is unchanged (for now) and genetically (pun-intended) she will be changed from that point forward, yet her identity is *already* unraveling, because in a generative light her past is brought into question—her birth, as well as the past before her birth.

Her ancestry and the constitution of her family are put into question. She will discover she is involved in things far broader than she ever realized. Identity, in this sense, is generative in that it always moves beyond our own histories. Our own narratives, which shape our identities, contain this element of historicity or sense of world history, which stretches back to before we were born and persists after we die—we are sons and daughters, as well as fathers and mothers, a link in a chain of generations.

The unraveling of Sarah's identity, and its ongoing reformation or re-constitution that is reflected in her lived-body, continues in her relationship with and the revelations of Helena. Not only is Sarah a clone, but she is womb-twin with Helena. Their personal histories have been completely separated, shown in their scars, habits and languages that they carry on and in their bodies. Helena shares the genetic "defect" that allows her also to be fertile, just like Sarah but unlike any of the other clones.

Similarities cannot help but be drawn, but the differences are stark. The philosophical question is: How can that even be said; how is there both similarity and difference simultaneously in identity? This cannot be a question of form or essence (A is A), because A is A and a bit of B too. We must look at the generative element present in identity, looking at the narrative of identity. Sarah and Helena are drawn into similarity and into difference, because they share generative elements of their narratives: womb, genetic predispositions, and so on. It is sharing that creates here the limits of what is *not* shared, that is, of what is different, of what allows distinctions to be made.

Individual difference can be addressed from the perspective of life-projects. Sarah and Helena, although sharing many commonalities (including being raised outside the experiment), are almost instantaneously noted for their differences. What shapes their identities and these differences are the projects they are seeking to complete, that is, their intentionalities or ways of life.

At first, Helena attempts to kill all the other clones, yet even as this project fades away and she becomes a part of the Clone Club family, her projects go beyond the concerns of the rest of the group—Sarah included. She wants to have a child and love her boyfriend, Jesse Towing. Sarah, on the other hand, has survival as her life-project, for herself as well as for her daughter, Kira. This is brought into the rest of the group, but for Sarah the focus grows more devoted to Kira, especially as the scientific interest in Kira grows. The other clones also have their projects. Nevertheless, while differing projects help to constitute different identities, they also produce the possibilities of commonality and shared projects.

The Clone Club can survive together and fill in for one another in their various projects (when various people impersonate Alison for example during the whole course of her campaign), which is one of the many devices played with in the show, yet it reveals the fact that they are a group and share elements of an identity even in difference.

I, You, We

In many ways, the Clone Club throughout *Orphan Black* functions as a tribe. There are rites of initiation of sorts (the giving of the phone, the introduction to the others, and the common experience of the unraveling of identity; this last "rite" is one of the reasons that Rachel will always be other than the tribe, as a self-aware clone she never experienced this unraveling), as well as fierce loyalty and sacrifice for the other members of the tribe.

There are individual members, whose individuality is asserted by their own problems and projects, for example, Helena's rehabilitation, Sarah's drive to obtain independence and security for Kira, Alison's desire for normalcy and the accoutrements thereof, and Cosima's health and love-life. These differing projects, with miniature projects contained therein, help constitute differing identities, even as commonalities are discovered and built upon, for example Sarah and Helena drawing closer in their perspectives on the world. Once again, the growth of a common identity (and a personal identity) and of a differing identity can be understood generatively, a continuous stream reformatting and reaffirming or redefining itself.

The Danish philosopher Søren Kierkegaard stated that life must be lived forwards, but understood backwards. What *Orphan Black* demonstrates is that this must be broader, and this is why an analysis of identity must consider the generative conditions (those of world-history or of intergenerational projects) of identity, and not merely the genetic, which only includes our own personal history, which for Kierkegaard would be the beginning, the point to look back on. The need to consider these elements is evident in looking at the differences between Leda and Castor.

Although both are originally sourced from the same person, who is genetically chimeric, the projects of Leda and Castor have intentionally different aims: Leda, to explore the science of cloning and the possibilities therein; Castor, to explore the possibility of creating an army of clones. While the clones learn this in the progression of the narrative, the revelations about the past form particular futures for them, including the need to find a cure for their genetic defects. That is, life may be lived forward, but to fully understand it, I must go back before the beginning of my individual life. We must acknowledge that identity is wrapped up in the intergenerational projects of the species. It is then, as that life is lived, that the imagined futures we wish to realize help us interpret the pasts we bear and the presents we share. As we understand our own identities generatively, we acknowledge that identity moves beyond our own deaths, insofar as we contribute to the whole ongoing project of humanity, of the "life-world" we share, that is, the common elements and conditions that we help to shape and pass on.

Leda and Castor are not different and similar just in terms of their history—of their founding and origin. The Clone Club and the Castor Project, as they interact (both in alliances and in competition), discover their own similarities—fears, anxieties, origins—and differences. These similarities and differences create both internal and external limits; the members of Clone Club in discovering and learning more about Castor recognize their own similarities among themselves more (thus creating a stronger tribal identity) and the differences between their tribe and the Castor tribe.

Husserl calls this the issue of the "homeworld" and the "alienworld." My homeworld is the place of meaningful relationships and connections where my identity is solidified in opposition to an alienworld, which is a place where there are no anchor points for me to understand my place in the world. In short, it is by encountering difference that the understanding of what is *home* becomes constituted more fully. People who travel come to understand this, and it is dramatically presented in the Clone Club's (particularly Sarah's) reactions to the Castor clones.

One of the final scenes in Season Three, when they all gather around the table together in Alison's shop, is only made possible because of the identity constituted in opposition to the identity of Castor, whom we cannot imagine enacting a similar scene. The homeworld is defined, because something other and alien has been brought up against it. Yet even there, because of the generative, ever-evolving and reinterpreting of identity, there are moments where the home and the alien touch and share, for example, when Helena kills Rudy in the garage. He is already dying when he encounters Helena, but as he bleeds out, Helena mirrors his position on the ground, not because of her own psychosis, but because she sees parts of herself and of her sisters in the man dying on the floor (although she attempts to deny their similarities when Rudy points it out, still affirming an identity defined by difference).

The complex relation between the self and the "we," which is in play in the descriptions above, needs to be more fully explained. The "I" does not merely belong to a group, but co-constitutes itself with its homeworld; the Clone Club is only possible, as the members, the individuals, constitute it together. The phenomenological truth here is put forward by Husserl simply: the I and the world are co-original. There is no world without the I, and there is no I without the world. The I is present in the world, in the lived-body, which therefore places every "I" in a context, in a terrain, in a homeworld, which is constituted by "I"s that have come before and "I"s which will inherit it. I am a link in a chain, but I am *a link*, an individual as well as a member of a group that helps me define who I am.

Unknown Self and World

Identity is not an individual affair. Identity itself is intersubjective—it is constituted among others and others help to constitute it. As such, there are elements of identity which are unknown to the "I." For Sarah and all the other "unaware" clones, this is painfully obvious, especially when they become aware.

No "I" is born into the world fully aware of itself and of everything else in the world. Others must introduce me to myself and to the world. Phenomenologically, we could say that while the world will only be brought into existence by being for an "I," the "I" is only brought into itself by being there in the world. This is understandable if identity and the I are taken to be generative, ever growing, re-interpreting, and re-constituting themselves and the world around them. This "self-awareness" is an ongoing project, with elements concealed and obscured by time, by perspective. This is true not only for Sarah and the Clones, but also for every individual.

Self-awareness, though implied in every act of consciousness (*I am* thinking), is an ongoing project in terms of its shared elements. In the course of living my life, I develop relationships and an awareness of my self, because of those relationships. This is the way that I understand myself and the world around me. This is Sarah's journey in *Orphan Black*, from the punk on the train platform being denied access to her child, to the self-confident young woman at the end of Season Three, eating and toasting with her family and getting to look at Kira in the snow, Kira whom she has protected and for whom she has sacrificed.

With all these ideas in mind, let's look at the opening scene again, where Sarah and Beth meet, when the generative journey of identity makes a turn in Sarah's life. Looking backward, we know how alone Beth was; we now know the odds she was up against, what damage and hurt she had already endured when she met Sarah on that platform. She was an individual wrapped up in a homeworld, which was changing dramatically and was far more sinister than she realized before she was a part of the Clone Club. Beth's identity had been unmade and its re-constitution was half-baked, as it were.

Beth could not endure, so she decided to end her life. As she prepared to take this act, she turned around and saw another one, one that she had not seen a record of nor had met. In meeting her, she knows that she has just destroyed that other person's identity and world. What can she do? Nothing, except hope that after her death, the other one dismisses the surreal experience, considering it a drug-induced hallucination or some sort of weird coincidence.

But that will not happen; the Rubicon has been crossed for this other clone, this young woman who looks so harassed and lonely. Good luck, my other self, she might think to herself, good luck and Godspeed. The journey begins for you, as it ends for me. May it be more kind to you.

Sarah watches Beth take a few steps to her death, and it all begins for her. Identity is thrown open; so many questions are posed to her narrative. The appearance of her world has shifted and will continue to shift. Past experience is not disregarded, but it is reinterpreted and rediscovered, even as much of it remains hidden. In discovering her own place in a larger narrative, in one that is generative, stretching beyond her own life and life-world, she values that which will endure past her own—and the child which she was trying to reach on the phone becomes the most important element, and in the end, all she wants is to meet her eyes and to let her know that she is not alone, that the story has come before and will go on.

What Does It Mean to Be Me?

How does identity persist through change? Philosophers, poets, theorists, and theologians have wondered about this since the days of ancient Greece. How can we say *this* is the same Socrates, both at age five and at ninety-five? How does the "form" or essence of Socrates remain the same throughout all the changes? Most philosophers have wondered whether Socrates's identity is a single substance or thing. What is it that *remains the same* throughout an individual life? Others have wondered how change is even possible, and how there can be an identity that goes beyond an individual life, from the past and into the future.

As the world shifts and the science fiction behind *Orphan Black* becomes less and less fiction, the issues and problems raised by the story need philosophical attention. Yet even if science or politics were to stand in the way of human cloning, the issues are still there. What does it mean to be me? Where am I going? Where have I come from? Where do I belong? These all-too-human questions are the ones posed by Sarah Manning and the other characters, cloned or not.

Beginning from our own lived-bodies, we face the world and the history we belong to in ways that constitute us and our identity, but also in ways that allow us to change what we see and what will be. In the end, identity is something that can be shared. Yet it is that very commonality, that sharing, that draws the limits around the individual.

I am like you, and I am not like you. We are variations on a theme, all of us, in different ways.

PART III

"You want to grow a tail, that's your business."

8
Laughing in the Face of the Absurd

ROB LUZECKY AND CHARLENE ELSBY

"At any streetcorner the feeling of absurdity can strike any man in the face," writes Albert Camus in *The Myth of Sisyphus*. For Sarah Manning, that feeling strikes on a subway platform, staring at her own face as the woman attached to it jumps in front of a train.

Just after the incident, Sarah investigates, only to find that the face she always thought to be her own also belonged to Beth Childs. Sarah, at this moment, becomes aware of another self who is not herself, and this is absurd in the sense of being impossible. This is a moment of existential awakening for Sarah; it's the beginning of the existential turn, the forced expulsion from a life of habit and routine, the sudden realization that a long-held proposition can no longer be held; it is no longer true that *everyone has their own face*.

How do you deal with such a realization? It is, in principle, reversible; it's completely possible, as Camus says, for someone to experience such an awakening, and then just continue living as if nothing had happened. They may ignore the consequences of the awakening and gradually return to the mechanical life, clinging to the comfort of a supposedly purposeful universe and ignoring the truth of the absurd. It would be easy just to say that Beth was Sarah's twin or double, that there's nothing especially interesting about that, and go back to life as it was. While this would be one way to avoid existential anguish, there are, according to Simone De Beauvoir (as she explains in the second part of her *Ethics of Ambiguity*), several ways to do this.

For Beauvoir, Sarah, prior to her existential awakening, is probably living as what she calls a "subman." The subman lives as if they are a child, in the sense that the subman lives in a world that is already constructed for them, never questioning the authority of the system in which they are happy to find a place for themselves. Their goal in life might be to become a decent accountant in an established firm that offers good benefits. The subman never has to deal with an existential awakening, because they are never awakened from the mechanical life. Everything simply *is*, and it is always assumed that however it *is, is best*.

To remain as a subman, however, proves extremely difficult, given the level of self-awareness that humanity has attained. At some point, an individual will be faced with some kind of inconsistency in the system to which they have become accustomed. They must either explain this inconsistency away as somehow consistent with the system to which they hold dear, or they must deal with the realization that the system is flawed. To the existentialist, the latter option is more honest, as it seems incontrovertible that *the system is flawed.*

In Sarah's case, the system is so flawed as to allow for experimental children and the patenting of genetic material. And if the system is flawed in one way, then surely there must be other ways; soon the individual, on their way to becoming existentialist, recognizes that the constructs with which they have become so familiar are simply human constructs, that there is nothing on which to base their value than the arbitrary decisions of other subjects and their whims—that it may very well be the case that the whole reason for your existence is the ethically suspect action of Dr. Aldous Leekie.

Camus refers to Sartre's concept of "nausea" to explain the feeling you get when you're suddenly self-aware, but still living in a world of submen. Other people, behaving mechanically, living by routine, appear as machines, as something distinctly *inhuman*. Camus defines "nausea" as "discomfort in the face of man's own inhumanity." The concept itself points to the absurd—the idea of an inhuman human is internally contradictory. There should be no possible way to encounter an inhuman human, *and yet we do*.

There do exist humans who behave mechanically, who have no qualms about engaging in a conspiracy to continue a covert experiment on unsuspecting human subjects. To the awakened existentialist, the world of the unawakened seems patently absurd. This world starts to seem unfamiliar, and where familiarity is comfortable, unfamiliarity is uncomfortable. Even the people who don't have to behave mechanically in order to conduct brutal experiments start to seem silly, those people constantly worried over problems like paint colors and whether the subway train will be on time, the kinds of concerns that, to the newly awakened existentialist, become completely overshadowed by the realization that *the system is flawed*; *not everyone has their own face*.

For Sarah, the good existentialist, it is no longer possible to go about her business without having to consider the consequences of this new realization. At this point, her options for how to deal with the future are limited to those that account for this new level of awareness—the awareness of other selves.

Beth's Death and the Death of Meaning

Sarah's experience with Beth Childs is the closest thing that we can experience to our own death. When Sarah is confronted with the stark reality of Beth's death, it is like she is being forced to confront her own death. But wait a minute, you might object—Sarah's watching Beth die is not at all the same as experiencing her own death. If only this were true. Though Beth is a stranger to Sarah, we cannot say that her face is new, for it is the same face that Sarah has confronted everyday of her adult life.

But faces are not all that define us. Materialists and various other Dyad employees would argue that we're nothing more than our DNA. If, for a moment, we accept the claim that what we are is essentially our DNA, and we recognize that Sarah and Beth share the same DNA, then the person who Sarah watches die on that dimly lit platform is herself. In this sense, Sarah is not merely bearing witness to a stranger's death. When Beth steps off the platform, it's the same as if Sarah is stepping off the platform. Sarah is participating in Beth's act of dying. In a sense, every time one of her sisters dies, Sarah dies too. Beth's death is Sarah's death, and so is Katja Obinger's, Jennifer Fitzsimmons's, Ania Kaminska's, as well as the deaths of the women from Helsinki and the Polish clone.

Maybe Sarah's repeated experience of her own death is not such a bad thing. After all, when we first meet Sarah, we gather very quickly that her life is somewhere on the south side of ideal. She's certainly not living the seemingly posh life of Rachel, and she doesn't even enjoy the outwardly perfect suburban life of Alison. Sarah seems to barely be able to hold down a job, and her life seems to be little more than a protracted struggle. Maybe Sarah's death at that moment is a type of reprieve, a moment of redemption when she is—finally—free of her life's tumult.

But Sarah is put in the even more absurd situation that, after experiencing her own death, she remains alive. If she were reprieved, then her death, far from being a moment of anguish, would be a celebratory moment of release. But instead Sarah must remain alive while experiencing her own death many times. Sarah's repeated experience of her death forces the existential awareness that in a very real sense, her moment of release is nothing meaningful. Sarah's death merely lasts an instant, quickly fading as she is plunged back into a world where nefarious organizations seem to be out to get her.

With death robbed of any meaning, so too the meaning of life disperses to nothingness. Shortly after Beth's death, Sarah comes to learn that her childhood was nothing but a series of lies compounded. It's not that there is one moment in life—like that fifth birthday party when there were no balloons—that is dubious. Every moment, from when we enter the world crying, to when we depart from it, is now cast into doubt. And it is the sort of doubt which cannot be resolved by the uncovering of a new fact or set of facts. When Sarah stands on that subway platform or in that police precinct, when Helena awakens in that prison cell, each is confronted with the same truth of their existence; each day brings with it new doubts, and nothing, not one moment is to be believed.

And this is the true force of Beth's death; it confounds the major philosophical categories and renders the distinctions between metaphysics, epistemology, and ethics rather pointless. When death confronts Sarah, it is apparently a metaphysical event, but it has diminished meaning, in the sense that it impossibly happens repeatedly, and death is supposed to be that life-defining event which only happens once. The awareness that death has no meaning is also a crushing awareness that we do not know how to comport ourselves to it, and there is no one and no institution that we can look to for moral instruction. When Beth falls off that subway platform, a hell of a lot falls with her; she dies and with her the possibility of meaning dies to the sound of an oncoming subway train.

Existential Ambiguity and Existential Absurdity

The existentialists believe that the human species faces, more than at any other period in its history, an increased level of self-awareness that lends itself to feelings of absurdity. Sarah Manning embodies the existentialist's anguish, as she becomes aware of the absurdity of the system in which she had previously lived, the futility of fighting against that system, and the ever-present sense of being a *one amongst many*, a subject in a world of other subjects who, though genetically identical, can never share her particular viewpoint.

While the existentialist may tend towards nihilism, or the feeling that in the face of such anguish one should just say, "F--- it," Sarah, along with Jean-Paul Sartre, Albert Camus, and Simone de Beauvoir, accept the absurdity of her situation and continues despite it, laughing in the face of the absurd.

According to Camus, when we say "It's absurd," we mean one of two things: either "It's impossible," or "It's contradictory." The idea of inhuman humans is contradictory. The idea that you're the product of an illegitimate cloning experiment and that your lover is hired help involved in a nefarious scheme to monitor the progress of this experiment appears impossible, until such time as it can't possibly be denied. It is definitely absurd.

Simone de Beauvoir claims that the current situation of humanity is paradoxical; man now recognizes the paradoxical situation in which he lives. Man is a subjectivity whose existence is limited to the present moment, which itself is a fleeting bit of time between the non-existence of the past and the non-existence of the future. She claims that it is evident in man's attempt to master the world, only to find that such actions are futile, giving the example of the atomic bomb—man masters the atomic bomb, and the atomic bomb destroys him.

Sarah Manning must feel more than most the paradoxical nature of her current condition. She's the result of an experiment that sought to control nature by determining her genetic makeup, and yet her freedom as a subject is evident as she seeks to destroy the system responsible for her origin. But from this perspective, Sarah is the bringer of absurdity, rather than the victim of it. Humanity comes into conflict with nature here because, despite Dyad's attempts to control the results of their experiment, Sarah, as a human subject, retains absolute freedom—a freedom which, according to the existentialist, she has a duty to exercise. As Sartre says, "Freedom is the foundation of all values" (*Existentialism Is a Humanism*). Sarah, faced with a world which she recognizes as artificial, exercises her responsibility for action; she exercises her freedom for the sake of everyone's freedom. When she decides in favor of her own freedom, she starts a battle with Dyad for the freedom of all humanity.

While an existential awakening might propel someone into despair and nihilism, Sarah avoids these pitfalls, creating a purpose to life that corresponds to the existential goal of willing freedom and destroying tyranny. Another key component of the existential lifestyle is that when someone makes a choice, they imply by that choice the value of whatever they are choosing as something good, and they make that choice on behalf of all of humanity.

While we may say that what's good for me is good for me, and what's good for someone else is what's good for someone else, the fact that I chose whatever I chose implies that I think that my choice is the better one. When Sarah chooses to fight the evil corporation, she implicitly chooses freedom for all humanity. The goal is not to gain freedom for herself and allow them to continue experimenting on others as long as they leave her alone; the fact that Sarah chooses to fight has a universal implication. If I fight, it means that I think everyone should fight. If I want to be free, it means I think everyone should want to be free. Sarah fights for all of us.

Imagine Sarah Happy

Sarah's plight mirrors that of Sisyphus, as conceived by Albert Camus in *The Myth of Sisyphus*. Sisyphus, so the story goes, was condemned to roll an immense rock up a hill for all eternity, only to do it again once it rolled back down. Sarah, like Camus's absurd man, recognizes the futility of her situation, and laughs in the face of it.

The existentialist awakening is a recognition of the absurd, which according to Simone de Beauvoir, involves a recognition of the futility of attempting to seek out absolute meaning. The good existentialist accepts the inevitable failure, and continues in their efforts despite it. As Camus notes, the tragedy of Sisyphus's torture is not the torture itself, but the fact of Sisyphus's awareness. The real torture is not a rock or a hill, but the futility of the action and his awareness of its futility. The existentialist's ultimate revenge on the purposeless world is to recognize its purposelessness and to *carry on*.

Camus says, "Happiness and the absurd are two sons of the same earth" (*The Myth of Sisyphus*). We must imagine Sisyphus happy because he has his rock and *his rock is his thing*. Similarly, Sarah is happy; her struggle with Dyad is her thing. When the word "absurd" is invoked in a colloquial sense, it includes not only the definitions given above—those of impossibility and contradiction—but it also has the connotation of amusement and ridiculousness. When the absurd arises, we laugh. What else is there to do?

The Season Two finale, "By Means Which Have Never Yet Been Tried," is a perfect example of laughing in the face of the absurd. Even as Sarah and her sisters deal with the trauma of living as human experiments oppressed by an evil corporation, there's room in the story for a clone dance party that flies in the face of Dyad and their evil plots. In the midst of a seemingly never-ending struggle for freedom and autonomy, the clone dance party is a big "F--- you" to the world and its purposelessness. While it would be easy to exist in anguish and despair, to truly embody the persistence of the human species as it exists in conflict with an absurd world, Sarah must continue to fight, and she must continue to dance.

9
Variations Under Ethics

RACHEL ROBISON-GREENE

Human beings value a wide range of things. You might value a well-cooked steak, country music, and outdoor sports, while I might value green vegetables, Netflix, and a good book.

There are some values that we share in common, and those commonalities are fundamental to the human experience. These shared values serve as a basis for moral systems and for systems of governments. Countries base their constitutions on systems of shared values.

One of the values that we all have in common is that, nearly universally, we value the right to bodily autonomy—the right to do whatever we want with our own bodies. You have a right (so long as it doesn't infringe on the rights of others) to go where you want, to do what you want, to eat what you want, to drink what you want, and so on.

Common restrictions on bodily autonomy are based on whether individuals in the community or the community at large will be hurt when you exercise your bodily autonomy. So your liberty and your bodily autonomy are restricted, in some limited way, by the moral responsibilities that you owe to other human beings. For example, though you have the right to smoke cigarettes, you don't have the right to smoke them indoors because of the harm that second-hand smoke does to other people.

The members of the Clone Club in *Orphan Black* face a more complex set of moral dilemmas when it comes to bodily autonomy. These issues are complex because of the relationship that each clone bears to the other clones and because of the role that each clone plays as a participant (albeit unwittingly or unwillingly), in a larger experiment. The moral waters are muddier in the *Orphan Black* universe.

Unconscious Selection

When we're introduced to Sarah Manning, she lives a lifestyle, which, for better or for worse, is autonomous. She doesn't seem to view society's customs or laws as particularly applicable to her. She does what she needs to do to get by. If selling drugs is what it will take to make a better life for herself and for her child, that's what she'll do. When she wants out of a romantic relationship, she leaves. If she needs some time to sort her life out, she uses her foster mother, Ms. S., as a constantly on-call babysitter, sometimes leaving Kira with her for weeks or months at a time with no advance notice about when she will return.

On that eventful evening when Sarah runs into the first of her sister clones, Beth, on the train platform, everything changes. Sarah does not yet know, but will soon find out, that there are about to be some new, serious constraints on her liberty and bodily autonomy.

It's a fact about Sarah's life, one that she is just starting to realize at this moment, that she is not like everyone else. Her world is one in which there are an undetermined number of people running around who look just like her. Right off the bat, Sarah takes advantage of this fact. She assumes Beth's identity. She lives in her apartment, works her job as a police officer, and sleeps with her boyfriend. Also, right from the outset, Sarah ignores the moral implications of her actions. She must feel that she's doing no harm—Beth is dead after all.

Sarah's life certainly does change that night. I would argue, however, that the moment that she runs into Beth does not constitute the most significant change in her life. Lots of people have doppelgangers. Sarah could have simply concluded that someone else in the world happened to look an awful lot like her, shrugged it off, and gone on living her life. The crucial moment isn't really even when she chooses to impersonate Beth in order to take money out of Beth's bank account to facilitate running away with Kira to start some better life elsewhere.

No, the crucial moment in Sarah's story is the moment when she has retrieved the $75,000 from Art. She is, in that moment, free to do whatever she wants. Free to start that new life that, up until this very moment, she so desperately wanted. She chooses not to do that. She chooses to give the money back to Alison (who had given it to Beth to investigate their situation to begin with). In this moment she chooses the Clone Club, and in doing so, makes some serious changes to her realm of moral obligations.

Endless Forms Most Beautiful

I have certain moral responsibilities to you, and you have moral responsibilities to me. Most of the time, however, what I do doesn't affect you very much and what you do doesn't affect me very much either. The situation is a little bit different when it comes to the members of the Clone Club. Their physical similarities give them the power to harm one another and to benefit one another in ways that you and I couldn't.

First, when people look like one another, they can be mistaken for one another. The clones exploit this fact to their benefit as often as they are able. Sarah's impression of Beth is just the first in a long series of clones pretending to be one another. This works out all right in some cases. It often gets the clones off of the hook. For example Sarah stands in for Alison at her block party when Alison is too drunk to deal. Cosima pretends to be Alison during Alison's campaign rally. Alison pretends to be Sarah for Sarah's meeting with Ms. S. and Kira in Season One ("Effects of External Conditions"). And, of course, Helena pretends to be Alison during the drug deal.

It can also be very bad to be mistaken for someone else. Anyone who has watched enough fictional crime shows on television knows that identical twins have much to fear from one another. Resembling someone else seems to be a free pass to commit crimes and get away with them. This is particularly true when cases are built almost exclusively on eyewitness testimony. It doesn't only happen on TV, though. In the 1950s, one member of a set of twins (the Finn twins) stole a C-46 transport plane and hid it in the Nevada desert. The prosecution was unable to obtain a conviction in the case because the eyewitness involved was unable to distinguish one twin from the other and was, therefore unable to rule out that the twin they were seeking a conviction against was actually innocent. In another case, a woman was raped and accused one twin of the crime, and it actually turned out to be the other twin (Brian Palmer, "Can Identical Twins Get Away with Murder?"). The situation can get messy, even in real life (though in real life, police seem to usually be aware that twins don't have the same fingerprints because environmental factors in the womb determine what your fingerprint pattern will be. The writers of *Orphan Black* don't seem to be familiar with this fact.)

This is a situation that Sarah finds herself in right off of the bat when she is in the unique position of being confused with both victim and killer when Helena kills Katja. As we can see, Helena can potentially hurt Sarah in a way that none of the rest of us can hurt one another. She can mess with Sarah's life in the sense that she might mistakenly be taken for the victim of a murder. This could hurt her in countless ways, in part because it might prevent her from receiving goods and services ordinarily provided by the government that they wouldn't provide to dead people. Of course, it would be an even bigger problem if Sarah were mistaken for a killer. She could face jail time and potentially even the death penalty.

Not all of the clones hang with the best crowds. This can raise the possibility of harms of different types for members of the clone club. Sarah's ex-boyfriend Vic, though relatively harmless himself, keeps company with dangerous drug dealers who are after him when he is not able to pay them back for the drugs that Sarah stole from him. So, when Vic meets Alison at rehab, he introduces into her world—and the world of her husband and children—a dangerous criminal element that they otherwise may never have encountered.

We are introduced to the clones when they are already in their late twenties or early thirties, so many of them are well along their path with regard to important life choices. Cosima is in a graduate program, Beth is a police officer, and Alison is married and is raising two young children. Each of these life choices involves maintaining a certain kind of reputation in the community. A reputation that a less discriminating clone—Helena for example—could easily damage if they behaved in the wrong way in public.

Because of this unique ability that the clones have to harm one another in profound ways, the balance that we strike between what they owe to each other on the one hand, and the freedom they should have to live in the way that they each, individually, see fit, will be different than the balance that the average person has to strike.

Given that the clones have this capacity to hurt one another, what principles should govern their actions? If they care about doing the morally right thing (which, granted, not all of them are going to care about), how should they live? As a first approximation, let's consider a principle like this: Act only and always in a way that you would want your clone to act, were they to find themselves in similar circumstances. This principle is sort of like the philosopher Immanuel Kant's Categorical Imperative, which says, "Act only on that maxim that you can, at the same time, without contradiction, will to be universal law." The difference, of course, is that in our case, we constructed our principle in a way that it applies to the particulars of the moral obligations of clones.

How would this principle work in practice? In Seasons One and Two, Alison has a drinking problem. As a result, she gets herself in some pretty embarrassing social situations.

She's plastered during the block party that she hosts at her house. Later, she ups her game and appears on stage in a community play so wasted that she falls off of the stage. Since this is a play that's open to the public, we can imagine that anyone from the community might have walked in to see that play—a thesis advisor of Cosima's, a child welfare agent taking note of Sarah's actions, or a co-worker of Beth's. When considering her obligations to the other clones, Alison must consider whether she would want them acting in a similar way in circumstances in which they might be confused for one another.

One implication this principle has for the clones is that they can't just do whatever they want. Our freedom is limited by the extent to which we harm one another by exercising it. Because the clones' actions carry with them greater potential for harm in both frequency and degree, the clones might be morally entitled to less freedom than you or I.

Variation Under Nature

Another question you might have considered when watching *Orphan Black* is whether or not it's morally acceptable for any of the clones to opt out of being members of the Clone Club. They are, obviously, morally blameless so long as they don't know they are clones. But once each clone finds out who she (or he, in the case of Tony) is, is he or she obligated to stick around? Because the clones are so similar to one another, are they morally obligated to inform one another about changes in their lives?

Tony is an interesting case because he has a male gender identity. It is clear, then, that despite the fact that the clones are genetic copies of one another, there exists pretty substantial variation between them. If the clones stay in contact with one another, they are in a better position to understand what kinds of things are genetically settled and what kinds of things are not. That's the kind of thing it would be crucially important that the other clones know about. Contact with one another makes each clone more self-aware.

Again, however, the requirement that the clones stick around and keep one another informed about changes in their lives constitutes a substantial restriction to their right to exercise autonomy. It's easy to imagine that many clones might be freaked out by their status as clones. They might want to have nothing to do with their brothers and sisters. But what they choose to do matters. Imagine, for example, that one of the clones contracts the illness that Cosima suffers from and manages to beat it for good? Shouldn't that clone let Cosima know what happened? What's the right way to balance autonomy rights with the moral duties that clones have in virtue of being clones?

Conditions of Existence

The clones also face very clone-specific problems when it comes to issues related to health and reproduction. The Leda project, of which the members of the clone club are a part is designed in a way that leaves the female clones infertile and also (and not by design) makes them vulnerable to a deadly disease. Strangely, two of the clones (who were twins in the womb together), Sarah and Helena, are not infertile. When the series begins, Sarah already has a five-year-old daughter, and by the time that season three comes to a close, Helena is pregnant with a child of her own.

The Proletheans drugged Helena and inseminated her, resulting in her pregnancy. Even though the conditions under which she became pregnant were inhumane and criminal, she is thrilled to be pregnant. She loves children.

Ordinarily, we understand reproductive choices as falling firmly under the rights we have in virtue of having bodily autonomy. A woman can choose to take steps to not become pregnant at all. She can choose an abortion or she can choose to carry the baby to term. She can choose to raise the child herself or she can choose to give the baby up for adoption.

Morally speaking, do clones like Helena and Sarah—the only members of the Leda project that can reproduce—have the same range of options as an ordinary woman making reproductive choices? Would it be morally permissible for Sarah or Helena to have an abortion if they found themselves pregnant but didn't want to be? The question here is not whether abortion is permissible or not, but, rather, do the clones have the same reproductive rights that non-clones do? They pretty clearly would be protected by the law in the same way that any other woman would be. But would they be doing something morally wrong to their sisters by choosing abortion when their sisters could learn so much from the pregnancy and from the ensuing child?

Relatedly, we ordinarily think that people have a right to privacy when it comes to their medical information. If you have a terminal illness, you're free to keep that to yourself and battle through it in private if you decide that is what is best for you. If you do choose to let people know about your private medical information, it is up to you *whom* you choose to tell. With the Clone Club, it's different. For any illness you have, it may be the case that your genetics have something to do with contracting the illness or with why your body is responding to the illness in the way that it is. As a result, it seems like it's your moral obligation to tell your sister clones your private medical information. Here again, we have a restriction of individual liberty and privacy.

Certain Agony of the Battlefield

Throughout the course of the series, we see at least two characters grapple with the idea of taking their own life. The suicide that we're exposed to first is Beth's. Beth could no longer cope with the pressures involved with being a member of the Clone Club. The choice of whether or not to go on living is, to be sure, a choice that uniquely expresses personal autonomy. But did Sarah do something wrong by expressing her autonomy in this way? Beth was a police officer, and, in that role, she was in a wonderful position to fix some of the most serious problems posed by a world with a bunch of clones running around in the same town. When she died, she left balls in the air (for example, the investigation into her officer-involved shooting) that might hurt her sisters if they fell in the wrong way. Did she have a moral obligation to stay alive, even if she was miserable?

In Season Three, we find Sarah and Helena trapped in the Castor camp. Helena manages to get away, and she finds herself in a room with one of the Castor clones. He's strapped to a chair with a surgical hat on. When she removes the hat, she sees that his head has been cut completely open and his brain is exposed and is hooked up to wires. Clearly, tests are being done on this Castor clone, and, what is worse, he is conscious for the whole process. He begs Helena to kill him, and she obliges, driving a scalpel through his heart, twisting it to make sure the job is done right.

Like the Leda clones, the Castor clones are susceptible to an illness because of the circumstances of their creation. Their illness is different—it is a brain illness that inevitably leads to seizures and, ultimately, to death. This is an unwanted side effect of the project, and the Castor project is working hard to understand it and to destroy it. They can learn much from a clone that is currently suffering from the disorder. The study of this particular Castor may do much to save his brothers. Is he morally obligated to allow the testing? Were Helena's actions morally acceptable? Again, the only way to answer this question is to figure out the appropriate balance between personal autonomy and what we owe to other people.

The Weight of the Combination

Can these moral considerations from *Orphan Black* tell us anything about morality in the real world? The *Orphan Black* universe is clearly science fiction, but I think we can learn moral lessons that deserve consideration in our own, albeit less exciting, universe.

We tend to think of our rights to bodily autonomy and liberty as being, at least in most cases, inviolable. People talk about individual liberty as if it is the most important of basic human rights. The considerations we've discussed here can motivate ourselves to ask, about our own lives, "What liberties must I give up because I have moral obligations to other people?"

In many cases, when we sacrifice liberties, it's because we have taken on role-based duties. I give up my liberty to go out and party hard every Friday night because I have a child and I have a moral obligation to care for that child. I give up my freedom to run off to some beach somewhere in the middle of the semester because of the role-based duties that I took on when I decided to become a teacher. We all give up liberties when we choose our roles. What this suggests, however, is that the free exercise of or liberty is secondary to the obligations that we owe others, at least in plenty of circumstances.

The story of *Orphan Black* is unusual in this way because the Ledas and the Castors didn't choose to be clones. This is something that they were forced into. Throughout the course of the show, however, they choose how seriously they want to take the relationships that they have with one another.

In the end, we root for the clones to be close to one another, and when clones value their individual autonomy over their obligations to the group, we tend to think they're behaving selfishly. Yet, at the same time, we feel for the clones when the show highlights the extent to which each clone's identity is defined by a project that they had no choice to opt out of. *Orphan Black* is therefore, a great illustration of the complexity of moral life.

10
How Can Clones Disagree?

AUDREY DELAMONT

What would *Orphan Black* be like to watch if our clones always got along, and never disagreed with each other? Pretty dull, I think. Imagine the time that would have been saved if Alison had just agreed with Cosima about telling Sarah that they were clones? If Sarah had listened to Katja in the car before she was shot, or if Helena wasn't ignored early on?

And *that's* just to mention things that happened in the first three episodes of Season One! Disagreement amongst the cast of clones largely drives much of the tension and drama that we all know and love. It's what makes us shout at our TV or computer screen saying, "You know Delphine is your monitor, Cosima! What are you doing?!" just like Sarah did.

Arguments. We've All Had Them

We've all been in arguments. I don't just mean screaming-at-the-top-of-your-lungs and slamming-things-down-on-the-coffee-table arguments. I mean more *sophisticated* arguments. These sorts of arguments usually stem from some sort of fundamental disagreement.

Now when this type of argument occurs, shouting the loudest just isn't enough to *win* (let me tell you, I've *tried*). You need to provide facts, arguments, reasons, and anything else you might have at your disposal to try and convince the other person that *you're* right and *they're* wrong, or at the very least, show them where they've gone off track in their reasoning. In philosophy, we call this *rational disagreement.*

If you've ever thrown up your hands in frustration because the person you're arguing with just isn't listening, or because they just *don't seem to understand*, you've already got a pretty good grasp of what rational disagreement *isn't*. But let me take just a moment to say what it *is*, so that we're all on the same page.

Rational Disagreement can only occur:

1. Between epistemic peers. What does an 'epistemic peer' mean? It means someone who is on the same page as you, intellectually speaking. Neither of you is smarter than the other when it comes to the thing you're disagreeing about; you're both equally good at thinking about it.

   Think about your reaction when a child disagrees with you about something, versus when a friend or co-worker does (for example, imagine how Sarah might feel when Kira disagrees with her and compare that to how she might feel if Felix disagreed with her). You probably don't take the child as seriously. Maybe they just don't understand *why* something is the case, or maybe they aren't even capable of understanding yet.

   Very likely with a friend or co-worker, the same isn't true—they do understand and *are* capable of following what you're saying. And so while you and the child may *disagree*, it isn't the same *kind* of disagreement as that which happens between you and your friend or co-worker. It isn't taken as seriously, in some respects.

2. When you both have the same evidence. *This* means that neither of you have an ace hiding up your sleeve, so to speak. You both know exactly what it is you're talking about, and know all the arguments for and against it.

3. When you disagree. This one should be obvious—it wouldn't be much of a rational disagreement if you didn't disagree!

One simple example of this sort of thing is disagreements between atheists and theists. Now keep in mind not just *any* argument between these two groups will count, but I'm sure that we can all imagine two reasonable people who are experts when it comes to arguments for and against God's existence. Both the theist and the atheist are aware of what the other thinks, and agree about all of the available evidence. They simply draw different conclusions.

Philosophers Disagree about Disagreeing

If two individuals are genuine epistemic peers and *if* they share all of the same evidence, then it seems difficult to understand just how it could be that they arrive at differing conclusions. For instance, if you and I are disagreeing about how much each of us owes after our dinner at a restaurant, it's very easy to figure out why we are disagreeing. In order for rational disagreement to be possible, it would almost be as if we agreed on each individual point ("If you had the burger, that's twelve dollars with tax, plus a soda which was four dollars . . .") but then somehow disagreed on what the final price was. Unless one of us is just really bad at basic arithmetic, we must be taking something else into account (some other piece of evidence) that makes us arrive at a different conclusion. How can it be that we disagree, despite sharing all the same evidence?

One answer is that, despite appearances, we don't *really* share all the same evidence. Here's what I mean. Has anyone ever asked you a 'Why' question, that you just couldn't answer? "Why do you like that color?", "Why didn't you like your food?", "Why do you love *Orphan Black*?" Sometimes we can answer: "Purple goes with my eyes," "I hate onions," or, "How can you even ask me that?! It's amazing!" but other times we just sort of shrug and can't articulate what our reasons actually are.

These sorts of reasons philosophers call *private evidence*. Private evidence can be almost anything, from gut-feelings, intuitions, or just the way things seem to you intellectually.

But some philosophers think that the idea of private evidence poses a big problem for rational disagreement. Remember that part of what it *is* to rationally disagree is to have all of your evidence shared. If private evidence can't be shared, then by definition, it can't be rational disagreement. However, if this private evidence—these feelings and seemings—don't *need* to be shared, then they don't even *count* as evidence, and don't matter.

Another explanation for rational disagreement is the idea of *starting points*. Your 'starting point' is almost like the lens through which you see the world. Take again the example of Cosima and Alison disagreeing about what to tell Sarah, early on in Season One. Alison, the soccer mom views the world through a very different lens than Cosima does. Cosima is interested in science and learning things, while Alison is interesting in keeping her family safe and not disrupting her suburban lifestyle. Both have access to the same set of facts (what they both know about clones, and what Sarah has pieced together herself), and yet because of their different 'starting points' they arrive at a different conclusion.

But not all philosophers are happy with this answer, either. And it's for almost the same reason as before. Either the 'starting point' makes an actual difference and should count as evidence, or it doesn't. If it does, then the evidence has to be shared. If it doesn't count as evidence, then it must not matter enough to make a difference.

So it seems that on one hand, rational disagreement seems to be fairly intuitive and make a lot of sense. But on the other hand, some philosophers don't think it's possible.

Clones to the Rescue!

Wouldn't it be great if we could test out some of these theories in a practical way? Thanks to our clones, we can! Think about it: our clones meet all the criteria!

1. They all seem to be epistemic peers. We know that biologically they are all 'wired' the same way, and despite the different academic choices they've made, each is clever enough in her own right to be on an intellectual par. Sure, Sarah can't compete with Cosima when it comes to 'evo-devo', and Alison knows more about figure-skating etiquette than Helena does, but on the whole, the gulf between them when it comes to everyday reasoning doesn't seem to be *that* great. As Alison says to Sarah, ". . . you're the only person I can talk to, and you're just another version of me!" ("Variation Under Nature").

2. The clones, for the most part, share the same evidence. Although in some key (and dramatic!) situations information *is* withheld from one or more of the clones, there are plenty of instances where everyone is on the same page.

3. The clones disagree! This one is obvious. Pick an episode, and one (or *all*) of the clones likely disagree with the others about *something*.

Our clones seem to be ideal candidates for proving that rational disagreement is possible! So let's look at three clone-related examples and see if they run into any of the problems that have been mentioned.

Hello Helena

For the majority of Seasons One and Two, the other clones don't take Helena seriously. From thinking that she's some weirdly biblical sociopath, to someone who wants to harm Kira, what Helena has to say doesn't strike the other clones as worthy of consideration. But were Cosima, Alison, and Sarah too quick to blow her off? If she is their epistemic peer by biological design, should they have given her arguments more weight, earlier on? Well, *no*.

Rule 1 of Clone Club might be that you don't talk about Clone Club, but Rule 1 of rational disagreement is that you have to be an epistemic peer.

But wait! Didn't we say that the clones were epistemic peers? We said that the clones *seem* to be epistemic peers given their biological framework. But *clearly* Helena is *not* on the same page as the other cast of clones. And I don't just mean that she disagrees with them (in the way Sarah might say tomayto but Rachel would say tomahto), Helena's upbringing seems to have placed her almost irreconcilably far away from anything that clones such as Sarah, Cosima, or Alison can understand. A life of stifling religious indoctrination combined with physical and mental abuse has left Helena in a position that none of the other clones can even *imagine*.

Perhaps we might attribute this to Helena having a different *starting-point*, rather than her not being an epistemic peer, but either way, it should be fairly clear that when Helena disagrees with the other clones, it is *anything* but rational.

Clones versus Rachel

Rachel Duncan, the primary antagonist of Season Two, has been self-aware since birth (or at least, as soon as she was old enough to understand). She routinely submits for testing, is on board with Neolution and even requests a new monitor once Daniel is killed. Despite her radically different upbringing, she seems epistemically *closer* to Cosima, Alison, and Sarah than Helena does. She's more rational, to say the least.

Do the disagreements the clones have with Rachel constitute rational disagreement? *No*. Rachel has a lot more information than Cosima, Sarah, Alison—or any of the other clones—combined. Despite the fact that Rachel seems to be an obvious epistemic peer in virtue of being a clone, any disagreements she partakes in with the other clones, cannot constitute a rational disagreement, due to the gulf between the evidence she has, and the evidence that they've managed to piece together.

Cosima versus Sarah

Not all rational disagreements need to be massive in scope. Let's look at a smaller, specific example:

Cosima: Right, well, I may have a monitor dilemma of my own. I'm new here this semester. I didn't bring anybody with me, but someone wants to be friends.

Sarah: Just stay away from them, Cos. Stick to the science, yeah?

Cosima: Stick to the science? What am I, the geek monkey, now?

Sarah: Seriously! Paul's a bad arse, okay? He's ex-military. If somebody's trying to get close to you, just stay away.

Cosima: Right, the old "Do as I say, not as I do"? ("Variations Under Domestication")

Sarah: Don't do it, Cosima. Stay away from Delphine, yeah?

Cosima: Well, if we're going to get past our monitors, we have to engage.

Sarah: Yeah, look where that got me.

Cosima:Cosima: Into bed with him?

Sarah: Okay, look where it got Alison, yeah? She smoked her husband with a golf club on a hunch that he was her monitor and, now, they're away at a couple's retreat, trying to repair their marriage.

Cosima: Okay, I'll obviously approach Delphine way more logically than that.

Sarah: No, you won't.

Cosima: Sarah, my situation is way different than yours is.

Sarah: Right, 'cause you're such a brilliant scientist.

Cosima: No, because she doesn't know that I know. I'm the one who's monitoring her. ("Parts Developed in an Unusual Manner")

What's going on here? Is this a case of real rational disagreement? The last criteria is certainly met: Sarah and Cosima *are* disagreeing. But does their disagreement meet the first two conditions?

To begin, we need to establish whether or not in this instance, Sarah and Cosima are epistemic peers. While Cosima's knowledge of PhD-level biology clearly exceeds Sarah's, when it comes to monitors, both seem to be on an equal intellectual playing field. Epistemic peerhood doesn't need to extend to all areas of your knowledge, just whatever is relevant to the disagreement. In this case, it seems that Sarah and Cosima *are* epistemic peers.

The second condition is whether or not Sarah and Cosima share all the same relevant evidence. It certainly seems like they do. Sarah has been forthcoming with her impersonation of Beth Childs with relation to Paul (Beth's monitor), down to some intimate details, as Cosima clearly mentions ("Into bed with him?"), but also the relevant ones as well, such as how much Sarah suspects Paul of knowing, what she thinks Paul is doing, inferences about Paul and Beth, and even the background information she knows about him. With regard to Cosima and Delphine, Cosima is upfront about the fact that she suspects Delphine immediately, and that she intends to monitor Delphine, while in turn being monitored by her.

That seems to be the extent of the evidence at hand: both clones are aware of the existence of monitors, and know that they will be or are being monitored. Both know that the monitors are typically highly trained (with Paul being ex-military), and both know that it is unclear whether or not they can or should trust their monitors. Having no prior experience manipulating monitors, we can only assume that Cosima's claim that she will be able to monitor Delphine, is just a supposition on her part. However given Cosima's highly intellectualized outlook on life, it isn't unreasonable for Sarah to agree that Cosima would be better at monitoring a monitor than, say, than Alison and her failed attempts at monitoring Donnie. So when it comes to shared evidence, both clones seem to genuinely know what all the facts are.

So! Sarah and Cosima are epistemic peers, they share the same evidence, and yet they disagree! But *why* do they disagree? Despite the fact that Cosima and Sarah have grown up in different circumstances, their starting-points do not seem to be radically different. Although their socioeconomic backgrounds are not completely the same—with Cosima being able to pursue a PhD, while Sarah is a con-woman—there is not such a gulf between them that they cannot empathize with the other's position (in contrast, consider how the clones view Helena: they aren't even able to imagine themselves in a position where the things she says and does make sense to them). Given their identical body structure, one could assume that Cosima and Sarah would have faced similar gender-related challenges as they grew, with each arriving into adulthood as attractive women, and sharing many other qualities.

So it doesn't seem as if their starting-point is what makes a difference. What about private evidence? Is there some piece of information that either Cosima or Sarah (or both) have, which subtly steers them towards disagreement? *Yes*. Both Sarah and Cosima know that fooling a monitor is hard—Sarah has expressed as much with her interactions with Paul. The difference is, Sarah *knows* what it's *like*. Despite the fact that both clones would likely agree that Sarah tried her best, tried to be convincing and did all that she could, only Sarah knows what that was *like*. That unarticulated feeling likely led to hesitation on Sarah's part—whereas the lack of it, likely led to confidence on Cosima's part.

Sounds pretty good, right? Well, remember the problem with private evidence? Philosophers who don't think rational disagreement is possible argue that private evidence either 1. makes an actual difference, or 2. doesn't make a difference at all. If it makes a difference in reasoning, then it should be counted as evidence, and thus needs to be shared in order to constitute rational disagreement. If it doesn't make a difference well then . . . *it doesn't make a difference* and shouldn't alter your reasoning.

So which is it here? Does Sarah's private evidence or feeling of *I don't think you can do it* and Cosima's *I think I can do it* count as *evidence*? *No*, and here's why. Imagine that you're a pretty good basketball player, and when you're at the free-throw line, you normally make your shot. Someone challenges you to shoot blindfolded. Someone else is going to shoot as well, and will be blindfolded also. They have about the same accuracy rate as you do. You both shoot, and, before removing your blindfold, you're told that *one* of you made the shot! Now, who do you think it was? Why *you* of course! Even though you both had about the same skill level, without evidence to the contrary (you didn't feel the ball slip out of your hands, the shot felt the same as all of the other times you've successfully made it), you're going to think that *you* were the one who was successful. It would be absurd to call that gut feeling that *you did it* evidence. It isn't. It's just a feeling, not grounded in anything rational (at least, not when compared to the fact that your opponent was just as likely to sink the shot as you were).

The same thing is happening here with Sarah and Cosima. Whatever private evidence they might have been experiencing, contrary to what some philosophers think, does *not* constitute evidence. Thus, Sarah and Cosima in this case, demonstrate a genuine case of rational disagreement!

Clones Save the Day

Well, there you have it. Despite the issues of clones and patents, of murders and suicides, of abducted children and bad hairstyles, at least we know it wasn't all for nothing. The clones have allowed the philosophers to rest easy in their armchairs, having demonstrated a case of rational disagreement.

PART IV

“You know I never would’ve got in if you’d said we were going to suburbia.”

11
Leda, Castor, and Their Families

CARMEN WRIGHT

Within the first few scenes of *Orphan Black*, the audience is made aware of family and its importance to the clone characters.

- **English clone Sarah Manning stole a large amount of drugs from her abusive ex-boyfriend in order to fund a new life for herself, her adopted brother Felix, and her daughter Kira.**
- **Soccer mom Alison Hendrix uses a gun as her primary method of protecting her husband, Donnie, and their adopted children Gemma and Oscar.**
- **Ukrainian assassin Helena has an attachment to Kira and is enthusiastic about carrying her own biological children.**

While there are many examples throughout the series that show that support and love of family are important for the main clone characters, none of them do so more than Helena's dream baby shower, which opens Season Three ("The Weight of This Combination"). Helena imagines that her relationship with the other clones has mended and they are celebrating the birth of her child. Alison organizes the elaborate baby shower, but brushes it off as simply "throwing it together." Sarah affectionately calls her "meathead." Kira is excited for Helena to have her own "monkey." Felix and Cosima present her with her food, and Felix highlights the closeness of their relationship by remarking that the food is "our sister's favorite." Cosima is free of the illness that has plagued her throughout Season Two. While each of the characters is a little out of character, this is how Helena imagines life with her sisters would be. The atmosphere is joyous, bright, and encapsulates the family camaraderie that could develop among the clones in an ideal normative familial structure.

Rather than relying on normative familial structures, *Orphan Black* draws out the complexities of both blood relations and chosen families. The show aligns the clones and their relationships with each other as a part of a blood family. These clone connections involve the Leda and Castor clones' blood relationship, Kira's blood relationship to Sarah (as the only child born from a clone), and Siobhan's relationship to her mother, who provides the genetic make-up for all the clones.

Ethan Duncan and Virginia Coady, Leda and Castor respectively, are not the clone's biological parents. The chosen families of *Orphan Black* build their groups through "adoptive processes." Siobhan adopted Sarah and Felix when they were young and acts as Kira's adoptive mother regardless of the child referring to her as "Mrs. S." Alison and Donnie adopted their children, Oscar and Gemma, when she could not conceive children. Ultimately, the blood and chosen family units created by the clones queer the normative family structure that does not appear in *Orphan Black*. Instead, they repeatedly create and re-create their own non-normative structures.

Turner's Uncanny Families

In her article "Clone Mothers and Others: Uncanny Families," Stephanie S. Turner discusses how Ellen Ripley of the science-fiction movie series *Alien* acts as a surrogate of women's agency as a cloned figure. Turner analyzes Ripley through each *Alien* movie from the last surviving member of the *Nostromo* crew to the genetically altered clone mother created by her former employer. From this progression, Ripley's status as a cloned figure questions sexual reproduction, its purpose within societal structures, and how it upsets normative family structures.

Turner writes that "the product of asexual, rather than sexual, reproduction, clones upset, or 'queer', the natural order, the sense of 'how things ought to be' in nature. The natural order depends on generational change, but clones, being genetically identical to their parents, do not contribute to that change. Existing outside of natural law, they are, in effect, biological outlaws." With this outlaw status, clones are no longer confined to the social constraints of marriage, kinship, and lineage, but also the meaning of family. These social constraints state that "people in family groups must be distinct enough from each other to fulfill their specific roles in the family, yet they must also be similar enough to each other to be considered an integral part of the family." Since clones exist outside the social constructs of family, they disrupt the definition of what constitutes a family. Turner notes that determining where a clone fits in the family structure is a difficulty if someone decides to clone themselves. Will this clone be a sibling or an offspring? This is evidenced by Siobhan Sadler and her relationship to the Project Leda clones. At first, she is simply connected through Sarah as her adopted mother. By the end of Season Three ("Insolvent Phantom of Tomorrow"), Siobhan is revealed to be the daughter of the Leda and Castor original, Kendall Malone. Now, it raises the question: Is Siobhan Sarah's sister since Sarah is a product of the cloning of Kendall's DNA or is she Sarah's daughter because Sarah shares the same DNA as Kendall?

It's Only Clones Tonight

As Turner predicted, the clones of *Orphan Black* are genetically related, but this does not mean that an automatic family group has formed. In the beginning, Alison is distrustful of Sarah, due to the latter's grifter past, and also Sarah's having adopted Beth's life. In "Variation Under Nature," Alison states that she feels very connected to Beth, does not believe Sarah can copy her, and that Beth was the one to teach Alison how to shoot a firearm.

In the previous episode, Cosima calls the group the inclusive name of Clone Club and in this episode, Felix refers to Alison as Sarah's sister. The inclusion of Felix into Clone Club is significant. The group, made up of the designated female Project Leda's clones, is selective with whom they tell their clone story. It takes multiple episodes and arousing suspicion from Paul and Art before Sarah is willing to reveal her clone lineage. With Felix, this is a different story because several minutes (on screen) after Sarah learns of her clone sisters, Felix is informed immediately, even if he was backup for Sarah ("Variation Under Nature").

Sarah and Felix frequently refer to each other as brother and sister despite their lack of blood relation and Kira calls Felix "Uncle." As the series progresses, Kira refers to each of the Leda clones as "Auntie." Alison accompanies Sarah, Felix, and Siobhan to the hospital when Kira is hit by a car where she explains to Felix that she feels like Kira is her own daughter ("Unconscious Selection"). Cosima's interactions with Kira come when the girl has been kidnapped by Rachel, and Cosima asks her to draw a picture for Sarah and teaches her a bit of science ("By Means Which Have Never Yet Been Tried"). Their interactions deepen when Cosima stays at Felix's loft when she becomes sick. Cosima reads *The Island of Doctor Moreau* to Kira and the girl guarantees her aunt is cared for. Their first scene together in Season Three, besides Helena's dream baby shower, is when they build a fort at the loft.

Kira's relationship with Helena is unique from the start. The girl is able to sense Helena from outside Siobhan's home and willingly follows the woman to an alley ("Entangled Bank"). Their quick exchange demonstrates a change in Helena's hostile demeanor:

KIRA: Where are we going?

HELENA: I'm taking you to meet someone [Helena's Prolethean handler, Tomas]. How can you be Sarah's daughter, child? How can that be?

KIRA: You're just like my mum.

HELENA: No, I'm not. She's not real. [*kneels*]

KIRA: Of course she is. Helena . . .

HELENA: Yes, angel?

KIRA: What happened to you?

HELENA: I don't know. [*they hug and Helena cries*] . . .

KIRA: I should go home now.

HELENA: Yes. Do you know the way?

KIRA: Of course.

HELENA: Good night, angel.

As the series progresses, Helena treats Kira like her own daughter and protects her from those who want to hurt her. Helena's own treatment toward other child characters exhibits her development. By Season Three, Helena is teaching Alison's daughter Gemma how to gouge someone's eyes out, and later on, she kills a group of drug dealers who threaten Gemma and Oscar's safety ("Insolvent Phantom of Tomorrow"). Even her relationship with Gracie, the Prolethean follower who is carrying more of her implanted embryos, improves once they have escaped the Prolethean farm. When Gracie suffers a miscarriage, she apologizes to Helena by saying "I'm sorry I lost *our* baby" ("Ruthless in Purpose, and Insidious in Method," my emphasis). In turn, Helena wants Gracie to be "auntie" to her baby. This conversation lays the groundwork for Helena to bring her confidant into Clone Club.

Helena refers to Sarah as "Sestra" since it was established they are twins and have the same birth mother ("Unconscious Selection"). When she mentions Alison or Cosima, she places the "Sestra" moniker before their names. In "Ipsa Scientia Potestas Est," Sarah teaches Helena about family:

> **Felix:** I'm sure I've got Ukrainian folk costume in here somewhere.
>
> **Helena:** [*hisses*]
>
> **Sarah:** Hey. Hey, hey, hey. You treat him with respect. You got it? That's my brother, which means he's one of our sisters. Family. You get it, meathead?
>
> **Helena:** Do not call me this.
>
> **Sarah:** Do you understand, Helena?
>
> **Helena:** He is sestra?
>
> **Felix:** Oh, God.
>
> **Sarah:** Exactly.

By the end of the episode, Helena dubs him "Brother sestra" and hesitates in killing Rachel to prevent Felix from serving jail time. In the Clone Club dance party scene in "By Means Which Have Never Yet Been Tried," Felix's inclusion with Kira (a biological relative to the Leda clones) solidifies his presence as the "brother sestra." He is not the outsider spouse like Donnie or Delphine; rather, he is a regular member of the group because he has been part of Clone Club since the beginning with Sarah referring to him as "the very best of us" ("Variable and Full of Perturbation").

The relationship between the Leda and Castor clones is filled with contention. Sarah uses "brother" in reference to the Castor clones when Cosima reveals that the Leda and Castor originals are siblings ("Formalized, Complex, and Costly"). Their relationship relies on being violent toward each other. The violence exhibited by Leda and Castor reaches its peak when Helena and Rudy fight in the Season

Three finale "History Yet to Be Written." Helena is instructed to eliminate Rudy by Sarah in order for the Leda clones to be free from Coady and her plans for them.

Helena tapes a screwdriver around one of her hands and tapes the other, whereas Rudy holds his knife. Helena declares the fight as "Prison rules. Only one of us leaves alive." When Helena is able to stab Rudy's bicep, he falls to the floor and is in the process of dying. It's appropriate that Helena and Rudy are the ones to fight each other, and not because Rudy is one of two remaining Castor clones. Throughout Season Three, Rudy has been willing to use violence to his advantage, such as when he threatens Kira in order to get information from Sarah ("Transitory Sacrifices of Crisis"). Helena is the only Leda clone with the ability to fight a military trained, Castor clone and potentially win. They are different, but closely related in their violence. Like her scene with Parsons in "Newer Elements of Our Defence," Helena comforts Rudy as he dies:

RUDY: Do you remember your childhood?

HELENA: Every minute.

RUDY: I remember sleeping. My brothers breathing in unison. We'd sleepwalk out of bed and pile in a corner, like puppies.

HELENA: When I was nine, I was made to shoot puppy.

RUDY: We're just like you, Helena.

HELENA: We're poisoned by men.

RUDY: We had a purpose. Just like you.

HELENA: No. You are rapist.

If there is any question as to Helena's violence, it is the dialogue in this scene. Helena's response to shooting the puppy compared with Rudy's description of the Castor clones is interesting, but also a warning to him that she has the disposition to kill them.

This Organism and Derivative Genetic Material is Restricted Intellectual Property

Among these queered clone families, there are the cloned mothers who present a threat to the natural, non-cloned world by agitating and reconstituting familial structures. Turner's argument for the reproductive roles of clones is multi-layered, but also instrumental to *Orphan Black*'s development of their clones. The basis of her argument originates in how clones are viewed as non-human because they are "experimental lab animals, patentable biotechnological products, and consumer commodities."

Since women act as the reproductive vessels, they are more invested in reproductive technologies in which their bodies are used for field tests with intrauterine ultrasounds and stem cells from umbilical cords. The familial units created by clone mothers has created an arena to question the reproductive agency of *Orphan Black*'s clones.

The Leda clones' role as reproductive commodities is demonstrative in the deformity in their genetic sequence. Their infertility triggers an autoimmune condition that causes polyps to grow in their uterus and lungs. With their genetic sterilization, the clones are unable to contribute to the societal family structures that are in place before their existence. Now, they are reliant on the family groups they have formed made up of the clones, their spouses, close friends, and colleagues. The infertility sequence warrants a change in the view of family structures and how they are able to navigate them. However, the infertility sequence does not apply to Sarah and Helena. Castor's Virginia Coady wishes to use Helena's fertility sequence as a way to cure the Castor clones of their neurological defect. However, this cannot come to pass since only the material from the Castor original can be used to manufacture a cure. In addition to the neurological defect, the Castor clones carry a sexually transmitted disease that can sterilize women. Coady's plans to weaponize the disease causes Gracie, the Prolethean carrying one of Helena's children, to miscarry when she has sex with her husband Mark, a Castor spy on the Prolethean farm. Leda's infertility sequence and Castor's sexually transmitted virus reflect Professor Duncan's plans for the clones to be produced without the purpose of reproduction ("Variable and Full of Perturbation").

The family created by Clone Club is considered non-normative by societal standards and they are a family by choice. However, this sense of family was not allowed to Helena and she becomes the perfect example of removed reproductive agency. Throughout her life, she's subjected to dysfunctional families, where choice was not an option. She was raised in a Ukrainian convent by nuns who abused her. At seven years old, the nuns said there were demons inside her and she was locked in a cellar. Eventually, she was able to pop a nun's eyes out in retaliation ("Ipsa Scientia Potestas Est").

Her involvement with the Old World Proletheans is one of dependency. The behavior of the Proletheans resembles that of the Dyad in the control they exert over the clones. The Proletheans act as a huge dysfunctional family, bordering on a cult, dominated by an abusive parent who violates the bounds as a father. She is ordered by her mentor, Tomas, to kill the clones because he tells her she is the original. The violence she exhibits, aided by her childhood in the convent, is fostered by Maggie Chen, the woman involved in Beth's shooting of a civilian, one of the factors in her suicide. It's only when Helena meets Sarah and Kira that she sees some semblance of a normative family. However, this is short-lived when the New World Proletheans bring her to their farm after she is severely injured in a fight with Sarah ("Governed by Sound Reason and True Religion").

At the farm, there is a difference of opinion between Old World member Tomas and New World member Henrik Johanssen, who was a former lab assistant at Project Leda. Johanssen believes Helena can conceive since she is the twin sister of Sarah, who was proved to be capable of conceiving and bearing her own child. However, Tomas believes that Helena is "defective and dangerous. Any child of hers would be a monster"

("Governed by Sound Reason and True Religion"). While recovering from her injuries, Johanssen "binds" Helena to him in a ceremony that strongly resembles a wedding and then takes her into a room ("Mingling Its Own Nature With It"). She is not aware of the egg harvesting procedure Johanssen has subjected her to, nor does she give consent in her barely conscious state.

She is able to escape the Prolethean camp, but returns when she is told her "babies" (her eggs) are back at the farm. In "Things Which Have Never Yet Been Done," Helena is injected with her embryos, but she is not aware of the technicalities of the procedure to harvest her eggs, nor does she fully understand that she is being artificially inseminated because her focus is on the ability to have children, motivated by her interaction with Kira. Johanssen fertilizes Helena's eggs so she can be the mother of a Prolethean child from a clone and implants the eggs into Helena and Gracie, his daughter, who could not consent to the procedure. Before the other eggs can be implanted into other Prolethean women, Helena takes the cryogenic tank containing her eggs and burns the camp ("Things Which Have Never Yet Been Done").

Helena's introduction into the show is by assassinating the Leda clones, her sisters. The violence exhibited by the other clones is justified as they are protecting their families from harm. The use of familial language to reference each other as "brother" and "sister" demonstrates how structured the clones believe themselves to be without participating in normative societal groups. The difference between blood and chosen families in the show is problematized further when taking into account that the blood families have removed the reproductive agency from the clones.

The message *Orphan Black* conveys about familial structures is simple: blood families are abusive and chosen families are full of love. The few examples of blood families are either violent (Kendall Malone murdered Siobhan's husband) or remove the child's agency (Henrik Johanssen implanting Gracie with Helena's eggs without their consent). The chosen families are met with favor even if they are violent in nature.

12
Not Why but Who

SARAH K. DONOVAN

Membership in clone club is nonconsensual and permanent. It might also get you killed.

Yet, under different circumstances the clones were regular people that you might have known. Alison could be that mom who hands out disapproving glares to everyone, everywhere. Sarah and Felix might have banged into you while dancing at a punk rock concert. You might have chatted about your day with Krystal while she did your nails. Now they are hunted by both the Proletheans and the Neolutionists, and trying to survive.

By the end of Season Three, the Leda clones seem to gain the upper hand when Sarah learns to ask "Who?" instead of "Why?" And the "Who?" of *Orphan Black* is not just Mrs. S.'s Mom, Kendall Malone (although she is a big "Who"), but many of the mothers and women of *Orphan Black*. As it tells us a familiar story about the psychological importance of mothers, *Orphan Black* leads us to look at what happens when women are viewed as less than human, and how they resist it.

If It's Not One Thing, It's Your Mother

Comic genius Robin Williams famously said "Freud: If it's not one thing, it's your mother." The Leda and Castor clones of *Orphan Black* would agree. Sigmund Freud (1856–1939) was the first psychoanalyst, and he discussed how mothers play a key role in human psychological and sexual development. Freud developed a talk therapy for his patients (think of the popular image of a patient on a couch, and a therapist in a chair, peering over the top of his glasses, taking notes).

It is now commonly held that Freud's view has serious problems, but it played an influential role in the history of ideas (particularly contemporary western European philosophy and feminist philosophy), so it helps us to understand how western society views mothers and women.

The Oedipus Complex is vital to Freud's view, and *Orphan Black* has lots of Oedipal themes. Freud assumes a heterosexual, married couple is raising a child as he unfolds the drama of the complex. As infants develop into children, they pass through three phases of development. First, infants are orally fixated: they suck on pacifiers, thumbs, or anything they can get near. Second, toddlers intensely focus on controlling their bowels (called the anal stage): they resist potty training, or they want to try out every bathroom they can find! Third, young children are focused on their genitals (called the phallic stage): they tug at their privates in public, without shame, and loudly repeat any words they have learned to describe their anatomy.

Freud is really interested in the phallic stage because it is here that the Oedipus complex develops (which makes children anxious but is key to becoming socialized) and resolves (which provides relief and maturity). Also, up until the Oedipus complex, Freud said that little girls act just like little boys. They also think that, like their male counterparts, they have penises (specifically, they mistake the clitoris for a penis). Enter: The Mother. Around age five, boys and girls love, love, *love* their mothers. They want their mother's undivided attention and are annoyed with their primary rival, their father, when he tries to interrupt the obsession.

Fearing Castration or Already Castrated

The male Oedipus Complex is straightforward. Little boys love two things more than anything else at this stage of development: their mothers and their own penises. A male child wants only to be with his mother. His father lays down the law, and says "No." The boy knows his father will punish him if he doesn't relent. But how will he be punished? By losing the only thing he loves as much as his mother. Clearly (okay, to a young child's mind), the boy must choose between his mother and his penis. This is Freud's famous fear of castration. It motivates the boy's ultimate choice to obey his father and find his own "mother" to love.

We can imagine that an unchecked Oedipal drama plays out with the Castor clones and Dr. Cody, their mother. For example, in Season Three, "Scarred by Many Past Frustrations," when Mark returns to his mother, she notices his wedding ring. He claims it was part of his cover with the Proletheans. She takes the ring off of him and says, "Just remember, it wasn't real. She is not one of us." Not only do these clones adore Dr. Cody (think of how, in this same episode, even the most dangerous Castor clone, Rudy, lovingly calls Dr. Cody "Mom" and displays affection towards her with a kiss on the cheek), she is the lead scientist in charge of the Castor experiment. She embodies what Freud calls "the Phallic" mother—domineering, intimidating, and ruthless (or, as if she had a penis, of which a phallus is the symbol).

Further, Dr. Cody controls the sexuality of the Castor clones (can you say "mother issues?"). She encourages them to have sex with many women (and consent is questionable, as Helena calls the dying Rudy a "rapist" in Season Three, "History Yet to be Written") and to collect hair samples from their conquests. The Castor clones bring home DNA samples, like trophies, to their mother. Freud would want to see them all, in his office, right away.

For Freud, the female Oedipal drama begins in a similar fashion but it is much more complex. Girls passionately love and hate their mothers. Sarah and Helena are examples of these conflicting emotions. While Sarah finds deep emotional fulfillment in meeting her biological mother, Amelia, Helena experiences murderous rage, and brutally kills her. Either way, the emotions are not subtle.

Freud describes the beginning of a girl's gender-appropriate Oedipus complex as a wake-up call about biology. Freud views girls as castrated or lesser versions of boys, and speculates on how mad they are when they discover this because, according to Freud, everyone (even if only secretly) wants to have a penis.

Girls are so mad, in fact, that they withdraw their love from their mothers and try to be the apple of their father's eye. This switch from loving their mothers to loving their fathers is how girls begin their Oedipus complex. For Freud, they will most likely not resolve their Oedipus complex. Why? Well, they can't be threatened with a castration that already happened! Their best hope at normalcy is to marry a man and have a son who will bring with him, albeit imperfectly, the penis that women always wished they had (I did warn you that this theory has problems). Freud thinks that the incomplete resolution of the complex keeps most women in an immature stage of emotional and intellectual development.

As gender-biased as this view is, we can imagine that Helena's relationship with Tomas is an example of an Oedipus complex gone wrong. Helena was raised in a convent and was clearly abused both by guardians in the convent and by Tomas, her father figure. Tomas created a fantasy world for Helena in which she was an original who was cloned. He fueled the fantasy with abuse and zealotry, and placed himself, as the authority and sole object of love, at the center. If this wasn't enough for Freud, Helena goes so far as to murder Amelia, and her clone sisters—she must be the one and only.

We can also place Rachel Duncan on Freud's couch. Rachel is completely traumatized by what she believed was the death of her mother (Susan Duncan) and father (Ethan Duncan). In multiple episodes we see Rachel watching videos of herself as a young child, with her parents, in what looks like an idyllic and undisturbed Oedipal triangle. While calm, cool, and calculating on the outside, inside she is a child in desperate need of parental love. We see a glimpse of this in Season Three, "The Weight of this Combination," when Rachel is trying to leverage her father, and Duncan poisons himself in front of her. In an uncharacteristic outpour of raw emotion, Rachel pleads with her dying father, "Don't leave me again." We witness a father respond with total rejection of his daughter as he says, "I am afraid you don't deserve me anymore." We can only wonder if Rachel's mother, revealed to be alive at the end of Season Three, can mend Rachel's shattered heart.

Castration Is Symbolic—Phew?

French psychoanalyst Jacques Lacan (1901–1981) saw himself as continuing the work of Freud, although whether this is true is debated. His theories (and writings!) are notoriously complex and open to interpretation. They have influenced thinkers from many diverse fields. Lacan was curious about how learning to speak a language, which also means entering into a complex linguistic and social order, influences the development of girls and boys into men and women. Following famous French linguist Ferdinand de Saussure (1857–1913) and influenced by renowned anthropologist Claude Levi-Strauss (1908–2009), Lacan thought about how language has an internal structure and social rules that shape how we understand the world. He called this structure the Symbolic Order, and this is central to his understanding of the Oedipus complex, and sexual difference.

Lacan thinks that the structure of language (so, not exclusively words themselves, but systems of signification or meaning) or the Symbolic Order, teaches girls and boys the social repercussions of sexual difference. Men are taught to be active subjects and women are taught to be passive objects. In this way, the Symbolic Order reflects the age old stereotype that men are rational subjects, and women are closer to nature because they menstruate, gestate, birth, and nurse babies. This stereotype is a kind of biological essentialism, or a belief that anatomy determines social behavior (so, a person with a uterus is hard wired to be overly emotional, and a person with a penis is destined to be more rational). It leads to a stereotypical division of labor between a masculine public sphere (of work, politics, power, and money) and a feminine private sphere (of home-life and family).

The genetic illnesses of the Castor and Leda clones symbolically reflect a version of a rational, active, male subject versus a life-giving, passive, female subject. The male clones's defect is neurological and affects their ability to reason, as evidenced by the tests on logical syllogisms that they fail when in the grips of neurological deterioration. It is also sexually transmitted so that they actively sterilize any women that they have sex with. The female clones' defect is reproductive and causes their ovaries to deteriorate so that they are unable to have children—except, of course, for Sarah Manning, who is so unique that the Neolutionists sought to surgically remove her ovary. The women are passive carriers of their illness and cannot transmit it sexually (as far as we know).

Like Freud, Lacan thinks that the Oedipus Complex is an important developmental moment. Boys love their mothers and want them to themselves. Boys know their fathers are their rivals, and they fear them. Unlike Freud, Lacan doesn't think boys are scared of literal castration. Rather, they understand this series of equivalences: having a penis = being male = having power as an adult = not being female. Boys want power, like their fathers. They fear a cultural or symbolic castration that would deny them masculine power. They worry that Dad will let them stay with Mom in the pre-cultural, natural world which promises limited access to power and culture.

Lacan names the mother's domain "the Imaginary Order" It is characterized by illusion and imagination, and all girls and women are destined by the Symbolic Order to remain there—with minimal power. In other words, in western culture, we live according to an inescapable bias that only men are the ones with real power who make the world go around. You might say, "But women can be in charge too!" and Lacan would not disagree. He would just say that they can, but they are acting like men when they do this. There is no place in our Symbolic Order for women to have power as women, and Lacan doesn't think this can change. While the Symbolic and

Imaginary are separate, Lacan recognizes that these orders are connected, as they are with the order of the Real, which completes his basic, three-part classification system. However, tension exists between all of the orders.

Mothers, Mothers, Everywhere but Not a One with Power

The feminist philosopher Luce Irigaray agrees with Lacan's theory that the current Symbolic Order supports age-old stereotypes where men are active subjects in the Symbolic Order and women are passive nurturers in the Imaginary. However, Irigaray, has different ideas about the ways in which real people construct language. Historically, men have had power, and women have not. It makes sense to Irigaray that men constructed language to reflect this. It also follows for her that language can be used to challenge these old stereotypes and construct new, more equitable,power dynamics between men and women.

Irigaray doesn't think that we can challenge the Symbolic Order until we really understand how, from a feminist perspective, it needs and uses the Imaginary. To do this, we need to think about what mothers mean to all of us (not your mom in particular, but our stereotypes about moms). "The mother," as a popular idea, is one of the most revered (Virgin Mary) and denigrated (portrayals of mothers as passive, asexual martyrs, or menopausal women who are no longer useful) symbols in western culture. Not all women are mothers, but mainstream culture holds this expectation, and often collapses being a woman with being a mother (and holds as suspicious or with pity women who do not want or cannot have children). Irigaray thinks that these types of ideas and associations keep women trapped in the Imaginary, without power. She believes that this is part of why women, and topics associated with them like birth, nurturing, and childhood, are excluded from intellectual history (and philosophy in particular). There is profound ambivalence associated with mothers (just think of Amelia!).

But Who Is the "Who" that We Seek?

Orphan Black has a lot of different kinds of mothers. We have mothers through adoption (Mrs. S., Alison, and Dr. Cody), mothers pregnant through intercourse (Sarah), mothers impregnated medically with biological material that is partly their own (Helena), and mothers impregnated medically with biological material that is not their own (Amelia, Gracie, and a whole host of other women who gestated the Leda and Castor clones). We also have the mother of the clones (Kendall Malone) who, in a genetic plot twist supplies both the female and male cell lines. Irigaray would appreciate this varied depiction of motherhood because it shakes up who we think a mother is (which in turn makes us question what role moms ought to play in society).

The moms of *Orphan Black* help us think about who controls women's reproduction, and the limits of that control. Irigaray is interested in those limits because it shows how a tightly constrained Symbolic Order cannot really capture the complexity of mothers and women—they are bursting out at the neat and tidy seams!

Control and its limits work in four different ways. First, the science is dependent on biological women for it to have any real application. The Duncans may have developed the genetic code behind the clones, but they cannot gestate them from embryo to viability in a lab. Only biological women can gestate and birth the clones. Even if the surrogates agreed to the pregnancy, most did not consent to be a part of a secret clone program. Amelia clearly expresses her dissent, and takes control of her own womb, by fleeing when she suspects bad intentions on the part of the surrogacy program.

Second, the Duncans tried to control the reproduction of the clones themselves when they manipulated their genetic code to produce sterility. However, the Duncans didn't intend for the engineered sterility to kill the clones—they lost control of the genetic code.

Third, Sarah Manning is a clone whose reproduction cannot be controlled. She's the only clone from the Leda and

Castor lines to produce a biological child, and the only woman not to be sterilized when exposed to Castor DNA (Season Three, "Certain Agony of the Battlefield"). Although we don't yet fully understand why, Sarah Manning is dangerous and vital to the clone project in equal parts, and she cannot be controlled.

Fourth, Mrs. S.'s mother, Kendall, is the mother of all of the clones and her DNA is special. She has a unique biological condition in which she absorbed her male twin in the womb, and has both female and male cell lines. She is the genetic point of origin, the Mother, and the "Who," of the Leda and Castor lines. She was a prisoner when Ethan Duncan harvested her DNA. When he could no longer control the project, he hid her identity from everyone, including his wife Susan. Kendall protected him in exchange for the right for her own daughter, Siobhan Sadler, to raise the only clone lost to the Dyad institute (or so they thought—they didn't know she was a twin), Sarah Manning. The fate of all of the clones hangs not with the original scientists, but on the original sister-mother of them all. And she's good at hiding.

The Mother of All Meanings

Gestating a human embryo from conception to viability is an important biological function that cannot be replicated in a lab. In this way, every person's birth can, in theory, be traced back to an existing person with a uterus. As significant as this empirical fact is, mothers also take on tremendous psychological significance. We see this both in how we revere and fear mothers and how the scientists and Proletheans of *Orphan Black* try to control mothers and women.

But why do moms get so deep under our skin and into our heads? For Irigaray, they are a haunting reminder of mortality and vulnerability. Men who want to feel powerful, masterful, and in charge are uncomfortable with such reminders. So uncomfortable that they might, like the Duncans or the Neolutionists, try to engineer birth and death itself. Irigaray thinks that Lacan's view of the Symbolic and Imaginary reflects men's desire to control and master.

While women and mothers obviously play a role in the Symbolic Order, that order works better when women and mothers are silent and unacknowledged contributors. They are mirrors or parrots who reflect back to men their masculine-centered vision of the world. They provide the support and encouragement for the public sphere. While this might seem depressing (for both women and men!) the good news is that this is not a permanent arrangement. Women do have their own subjectivity and culture. But they have to figure out how to create spaces for themselves within the Symbolic Order so that western culture can recognize them as real, active subjects in their own right.

Irigaray thinks that one way to do this is to underscore how women are often defined in relation to the role of mother, and how they defy this role by occupying different roles. She tries to highlight stereotypical views of women, and to subtly make fun of them, sometimes by amplifying them until they are ridiculous, so that we can see how made-up they are. This helps us to question grand narratives, like the Symbolic Order, that tell us that the situation we find ourselves in is just the way it is.

Irigaray mimics writings by Freud, Lacan, and western philosophers across the canon in order to show how they either stereotype women, or simply leave them out. By pointing to how stereotypes about men and women are constructed, she is already rewriting them. In three ways, the Leda clones of *Orphan Black* help us to think about what it means to question and rewrite, often in a playful way, stereotypes about women and mothers.

One, the Leda clones lay out for us a broad spectrum of how women may choose to live and express themselves—and we can, at first, neatly categorize them. Sarah Manning is the free-thinking rebel; Beth Childs is the neurotic depressive; Helena is the religious zealot; Alison Hendrix is the repressed rule-follower; Cosima Niehaus is the intellectual lesbian; Rachel Duncan is the calculating bitch; Tony is the street-wise gender-iconoclast; and Crystal is the gullible girl next-door. These clones highlight a rich segment of human identity. In this way, they demonstrate that women—even genetically identical clones—do not fit neatly into any one category. Irigaray would see this as playing around with sameness in a way that calls attention to difference (which is playful and disruptive to norms!).

Not only do the Leda clones show a spectrum of personality types, but they also evolve. Their character development reminds us that people are never static, and categories don't really fit any one person. For example, Sarah becomes more responsible to everyone around her, Helena learns to love both her sisters and her boyfriend, "Jesse's Towing," and Alison becomes less repressed and controlling (although, for most people, there are easier, more legal ways to do this). Even though categorizations and stereotypes about what it means to be a woman might try to lock identity down, the clones demonstrate that identity is constantly in flux. Rigid ways of classifying people may be descriptively real, but they are not immune to challenge.

Two, remember all of the impersonations of the clones by each other in *Orphan Black*? Sarah impersonated Beth and Rachel. Helena impersonated Beth, Sarah, and Alison. Cosima impersonated Alison. Alison impersonated Sarah. While impersonation serves many purposes within the series, it always has a side that is exciting, and sometimes even comedic. It was exciting when Sarah was impersonating Beth to colleagues like Art and Angie in Season One, or when, in Season Three, Sarah impersonated Rachel when Ferdinand visited the Neolutionists lab. In both situations, we worried that Sarah would break character and be found out.

It was comedic in Season Three, "Community of Dreadful Fear and Hate," when Cosima tried to impersonate Alison during her campaign speech to her small, suburban community. As she talked about family values, Cosima said, ". . . and, as a lesbian . . ." to which she quickly added "supporter." It was delightful to watch her question a heterosexual norm as Alison, who is a perfect example of a privileged, heterosexual housewife. Or when Helena impersonated Alison during the final drug deal of Season Three and couldn't at all pull off the role, and so, in a very perversely comic way, morphed into her usual psychotic self. We wanted her to right a clear wrong being done by the drug dealers when they threaten the Hendrix children (although we might question her methods, we always love Helena for defending children).

Whether we characterize impersonation as exciting or comedic, both ways underscore that identity is not stable. For what we find exciting or comedic is that they show the shifting identity of the characters themselves, and that out-of-character behavior can be funny (as it is in real life) and can challenge norms. This, again, demonstrates that it is fantastical to think that there is anything truly stable about people to be captured by something like a Symbolic Order. We can try to weed out contradiction in life and in language, but we won't be able to tell the stories of real people in this way.

Three, *Orphan Black* asks interesting questions about who is a *real* mother that subtly undermines stereotypes about mothers and women. Think of all of the mothers of *Orphan Black*! We have surrogates, biological, and adoptive mothers. As they mother side-by-side, their complexities as mothers are highlighted rather than erased. Mrs. S. *is* Sarah's mother, and Amelia *is* Sarah's mother. Sarah loves them both. Sarah *is* Kiera's mother but so *is* Mrs. S. Alison *is* a mother to her adopted children. Who *is* a *real* mother? Well, when we open up the category to include everyone, all of these women *are* mothers in important and real ways. This is the kind of flexible thinking and creative representation that Irigaray recommends in order to open up active roles for women as women in the Symbolic Order. "Mother" does not mean self-sacrificing object. "Mothers" can be active subjects.

The Question Is Not "Why" but "Who"

In "Certain Agony of the Battlefield," when Sarah is deathly ill after receiving a transfusion of Rudy's blood, she has a series of hallucinations that lead her to question "Who" instead of "Why." In her delirium, she and Beth are talking over tea and Beth reveals that she killed herself because she couldn't get her mind around being a clone. In admitting to a similar struggle, we get an intriguing exchange:

> **SARAH:** I don't understand the "Why."
>
> **BETH:** Why did you take over my life, Sarah?
>
> **SARAH:** For Kira.
>
> **BETH:** We do terrible things for the people we love. Stop asking "Why?" and start asking "Who?"

The hallucination ends with Beth softly touching Sarah's face, whispering "sister," and walking towards a light which snaps Sarah back to the moment of Beth's suicide. Sarah understands now that the key to understanding her origins is a "Who" not a "Why" and that people will do anything—anything—for the ones they love.

While Season Three concludes with us meeting the big "Who," Kendall Malone, many of the mothers and women of *Orphan Black* are also a big part of the "Who" in the first three seasons. Irigaray would see these diverse representations of women and mothers in active roles as positive responses to Freud's and Lacan's views about women's passivity, and as part of a project to break open a Symbolic Order that would relegate women, as mothers, to a silent Imaginary. But, if the live worm in Dr. Nealon's mouth, in "History Yet to Be Written," foreshadows anything, we can surely expect more on the "Who?" of *Orphan Black*.

13
Sisterhood's Back in *Orphan Black*

FERNANDO GABRIEL PAGNONI BERNS AND
EMILIANO AGUILAR

They are more than just a bunch of female clones. They are a club with rules. They are best friends and as such, they stand for each other. In *Orphan Black*, thanks to the development of cybernetics and technology, the sisterhood of the Seventies is back with a vengeance.

Within recent feminist discussion, Seventies feminism is frequently coded as passé, thanks, in part, to the 1980s backlash and the current postmodern movement—so dependent on girlness and youth—which typecasts feminism as uncool (Angela McRobbie, *The Aftermath of Feminism*).

Just when you think that feminism is out-of-date and therefore, as Sara Gamble puts it, "not worthy of serious consideration," a sci-fi TV show draws us back into the discussion of one of the most important prides of feminism: the concept of sisterhood. Sisterhood was the dream of a unified voice based upon the commonalty of women, a prevalent ideal that was championed by second-wave feminists during the Sixties and Seventies. It was close to utopia, but it worked for some time—until the whole project fell apart in the early Eighties.

Posthuman Orphans

Orphan Black is a story based upon the premise of the posthuman condition and body. The posthuman body is a mix of organism and biotechnology, a body made to surpass the boundaries of the normal, finite human body. Clones, as beings engendered within a laboratory, are posthuman in nature and not entirely human in the traditional sense.

To Rosi Braidotti, posthumanism introduces a shift in our thinking about what exactly is the basic unit of common reference for our species, what makes us human. This issue raises serious questions as to the very structures of our shared identity—as humans—amidst the increasing development of contemporary science. What Braidotti refers to as the "posthuman predicament" requires humans to think beyond their traditional limitations and embrace the risks of becoming-other-than-human beings.

Our "contemporary bio-genetic capitalism generates a global form of reactive mutual inter-dependence of all living organisms, including non-humans" (*The Posthuman*, p. 49). In this scenario, there is no strong distinction between humans and non-humans. Cloning is a good example: using animals as test cases and cloning them is now a legitimate scientific practice. In our times of advanced capitalism, says Braidotti, animals and humans alike have been turned into disposable bodies, inscribed into a global market of commercial exploitation (p. 70). It is rational, then, that clones are pejoratively called "animals" in Season Two's episode "Ipsa Scientia Potestas." Both, humans and animals are just subjects for experimentation in our posthuman world.

In *Orphan Black*, the female clones are the fruits of a series of illegal human cloning experiments, known as Project Leda, a scientific study from around 1977, the same decade in which feminism was in its strongest stage. Both, male and female are cloned to serve capitalist purposes such as the patenting of biogenetic procedures. Meanwhile, the men are cloned to serve in the military. The clones are not "originals", thus, losing their status as human beings. They are truly disposable, just experiments, heterogeneous mixes of organism and science. But (there is always a "but"), this form of advanced cold capitalism gives to the cloned women within *Orphan Black* the possibility of bringing back from the dead the ideal of sisterhood that second wave feminism had encouraged through the seventies.

The Birth and Death of Sisterhood

Second-wave feminism refers to a collection of feminist movements that emerged in the 1960s. Unlike the first wave, which was concerned with issues ranging from equal pay and women's full participation in social and political life, second-wave feminism proposed a whole revolution, a shifting of the social structures as the only path to woman's freedom. The second wave, indebted to the rise in political and student activism across Europe and America in the 1960s, was mainly concerned with eliminating gender inequality and the systematic oppression of women, and was focused on a range of issues including equal pay in the workplace, reproductive rights, and the ability for women to define their own sexuality (Hollows, *Feminism, Femininity, and Popular Culture*, pp. 3–4).

The body was the primary site for male oppression, since it was transformed into an object of male consumption and sexual desire and, because of that, a cultural sign of resistance. So, female desire and abortion were topics of great importance for feminists. And, as any casual viewer of *Orphan Black* knows, these are important topics within the series too. The first minutes of the episode "By Means which Have Never Yet Been Tried" is a compendium of feminist battles: Sarah Manning, the main character of the show, is heavily interrogated about her sexual orientation, number of sexual partners, contraception and abortion. The interrogation is, obviously, conducted by men, the ones who are always placing women under surveillance. *Orphan Black* reviews some of those feminist claims: Cosima, Sarah's nerd clone, is lesbian, while Sarah has had many sexual partners and an abortion. As Cosima screams in "Knowledge of Causes and Secret Motion of Things," in Season Two, "it's my body!", a cry that echoes the Seventies.

The two most significant branches of the second wave were radical feminism and liberal feminism. Radical feminists were deeply suspicious of any type of oppression and the social hierarchy; they pursued social transformations "through the creation of alternative non-hierarchical institutions and forms of organization intended to prefigure a utopian feminist society" ("Collective Identity," p. 173). Moreover, radical feminists championed the idea of sisterhood, which Renate Klein and Susan Hawthorne describe as "the recognition of a sense of political commitment to women as a social group" ("Reclaiming Sisterhood," p. 57).

The concept of sisterhood, however, proved to be problematic; by universalizing womanhood, it effectively eliminated differences of identity between women (Margaret Kamitsuka, p. 96). The "we," here standing for all women sharing a sole voice, implied a supposedly shared, quasi-universal female experience which "ignored the ways in which other factors, such as race, class, age, ethnicity, sexual, religious, or cultural differences, had not only historically established significant and often deeply divisive differences among women, but continued to do so" (Angelika Bammer, *Partial Visions*, p. 90) in the 1970s and early 1980s.

The first cracks within the union began to appear and differences, rather than community, prevailed. Sisterhood was based upon the idea that women were all alike because of their gender. But one day, a working-class woman stated that her position was different from all the others, mostly mid-class women. Later, a Latina, a woman of color or a lesbian followed her. Soon enough, those who were "different" went on to establish independent organizations, separate from the primarily white radical feminist groups that organized primarily along the axis of gender and soon, the concept of sisterhood started to dilute.

By the 1970s, second-wave feminists were subjected to a daily backlash at the hands of the media, which distrusted the revolutionary radicalization of the movement. Incredibly, feminists were depicted in popular media as lonely and depressed women due to the shortage of men in their lives! (Faludi, *Backlash*, p. 1). After the backlash of the 1980s, global unity entered a stage which marked the end of feminism and especially of global sisterhood in the 1990s and since.

Sisterhood's Back in *Orphan Black* (With A Little Help from Posthumanism)

Orphan Black brings back the original concept of sisterhood: Sarah Manning is a woman who in the first episodes of the show discovers that she is a clone rather than an "original" subject. On the "positive" side, Sarah discovers that she has many "sisters" scattered throughout the globe. These sisters are, in fact, more clones of the same DNA from which Sarah has been raised.

As the episodes go by, Sarah and her sisters form a bond which resembles the sisterhood of the 1970s. Even with their many differences (of class, of sexuality and nationhood), the different clones bond together to battle oppression, here metaphorized in those who want to imprison or kill them all as a way to interrupt this confederacy of female clones. As can be seen, years have passed since the Seventies but society is still trying to interrupt or backlash sisterhood. Even the original donor is involved in male oppression. In "Insolvent Phantom of Tomorrow" audiences learn that the source material and original donor for both the female and the male clones was Kendall Malone, a woman who absorbed a male twin in the womb during gestation. This unusual trait meant that she carried both male and female cell lines. It must be observed that Kendall is a gender-neutral name, which highlights the arbitrary use of language to designate gender.

This recuperation of sisterhood for a new millennium is possible only thanks to the philosophy of posthumanism. Only with the transcendence of the natural boundaries of humanity that posthumanism proposes, the philosophy of feminism and sisterhood can be raised from the ashes.

Feminism as a philosophy rests on the assumption that what we used to call "the universal subject of knowledge" is a falsely generalized standpoint. The social discourses and general assumptions that govern the production of knowledge and our daily practices, even the simplest, tacitly "imply a subject that is male (and also white, middle-class, and heterosexual) as Braidotti has argued (*Nomadic Subjects*, p. 98). We, society, understand the universal as the lighthouse from which everyone is judged. The universal is the white, heterosexual, Christian, Western man. Everyone who does not fit into this category is an Other. Now, let's be sincere here. How many people in the world fit within this category? Is it the majority? If we include women, people of color, and Asian, we find that this supposed universal is, in fact, a minority. It is just a social construction so white, heterosexual, Christian, Western men can have the upper hand.

Feminism proposes the replacement of this "universal" subject with one that is informed by other variables, such as sexual difference, ethnicity, class, or race. Thus, what we once saw as "the universal" appears as a very particular approach. "This particularity also explains its power of exclusion over categories of people who are deemed 'minorities', or 'others'."

Many of the issues raised by the ideal of the supposedly universal subject, are criticized in *Orphan Black*. Also, the series foregrounds the return of sisterhood. But it does so astutely, avoiding and reviewing some of the problems that sisterhood faced in the past decades. If traditional sisterhood encouraged homogeneity, *Orphan Black* celebrates difference. Yes, even if they are all clones of the same matrix. As Cosima, the nerd clone says in episode ten of Season Two: "God, we're so different."

We Are All Alike but Still We Are All Different

First, the series is critical, particularly through the multiple characterizations of Tatiana Maslany, about the possibility of a universal model of female body: the physical appearance of the clones put under discussion the ideal of female beauty still tinted by European's ethnocentricity. Sarah embodies the contemporary grunge American woman who wears loose, disheveled clothes. Cosima has glasses and braided hair, Helena has an animalistic appearance, one almost emptied of any trace of femininity (makeup helps to build the image of a pale, slightly crazy woman, with dark circles around her eyes but with a hidden, latent and savage beauty), and Alison represents an uptight American housewife. Thus, and unlike the sisterhood of the Seventies, the series enhances the idea of difference (of class, social background, personality, sexual orientation, nationality) even when they all share the same DNA. They were all alike, but are also different and the difference is here not downplayed.

The clones in *Orphan Black* also have differences in some portions of their genome. These portions have certain encoded messages which translate into the clones' tag number, a mark of their difference, used to tell them apart. If superficially identical, they are genetically different and their appearance and personal histories, together with their social background, place them as different from one another. Thus, these women do not fall into the trap of 1970s feminism: "partly in an attempt to achieve political consensus, feminists have often assumed a universal female body, an assumption that has usually left some women silenced, inhabiting the borderlands" (*Writing on the Body*, p. 3).

One of the pillars of our current times is, according to Braidotti, the post-secularity of posthumanism. According to Braidotti, we should think about secularism in terms of polarization between religion and citizenship: in the past, women were assigned to the private sphere and their participation was encouraged in religious or dogmatic matters, while external social activity was reserved only for men. In *Orphan Black,* women participate more actively in many areas; so, we have senior executives clones (Rachel Duncan), independent, self-sufficient and hard-as-nails women (Siobhan -aka S-), a scientist (Cosima), and even a doctor carrying a high military rank (Dr. Virginia Coady). These different jobs are also markers of differences among women.

The similarities between Sarah and Beth trigger the confusion that kicks off this game of mirrors and initiates the connection between the different women. It is precisely this physical empathy that holds together, at first, the notion of sisterhood throughout the series. In other words, while their collective identity is constructed upon the physical resemblance (because of the cloning project that ties them together in their posthumanity), this group intends to actually strengthen their ties. They are interconnected at first because of their gender and genetic code, which evoke new patterns of interconnectedness and affinity. Soon, their sisterhood expands thanks to new forms of affinity: real female friendship.

A New, Inclusive Sisterhood for the New Millennium

Sisterhood, as understood in the Seventies, is truly dead when *Orphan Black* begins. Women are not there to help each other anymore. In the episode "Variation under Nature" from Season One, Sarah, passing for her cop "sister" Beth, tries to adjust her duty belt and gun, without any luck: she simply does not know how to do it. A female cop offers her help (that Sarah is grateful to accept), but the help was just a gesture of mockery: the woman retires without giving any aid and with disdain in her face. This scene, together with the initial complete distrust among the clones at the beginning of the same episode when meeting each other, formulates a scenario of total individualism among women.

As the series progresses and the clones begin to trust each other, friendship comes to replace individualism. Each girl stands for her sister, resembling straightforwardly the ideal of sisterhood from the Seventies. Alison, the clone who puts more effort into building walls between the sisters, is also the one who makes the first gesture of sisterhood in *Orphan Black*: she passes for Sarah in the episode "Effects of External Conditions" to help Sarah retain custody of Sarah's little daughter.

This Clone Club is characterized by collective action. An example occurs in "Insolvent Phantom of Tomorrow" when

Helena risks her life by facing a gang of drug traffickers in order to help Alison and her husband Donnie (Kristian Bruun). On many occasions, one of the clones impersonates another to "get her out of trouble," usually in a race against time, as happens with Cosima and Alison on the eve of the elections at the school, or when Sarah impersonates Alison in "Knowledge of Causes and Secret Motion of Things."

It's essential, says Braidotti, to dismantle the notion of "pejorative otherness" as it is manifested by traditional theories. Braidotti notes that there are processes that operate negatively on marginalized people, creating partiality and ignorance about those considered as "others." This new kind of sisterhood for the new millennium creates bonds supported, unlike traditional sisterhood, by the acceptance of those others. *Orphan Black* sweeps away forms of ill-founded knowledge and prejudice and suggests to the viewer identification with this sisterhood.

The supposed "first rule" of Clone Club is that outsiders are not welcome, a situation that resembles traditional sisterhood. Sisterhood was very suspicious of those considered as not-feminists, especially men, because they were viewed as part of the oppression.

However, this postmodern version of sisterhood avoids making the same mistakes as its counterpart in the Seventies. In the first episodes of the Season One, Sarah has the help of three men: her friend Felix (Jordan Gavaris), Beth's cop partner Art (Kevin Hanchard), and Raj (Raymond Ablack) a Hindu who fixes the computers within the police department. Neither of them fit within the ideal of universalism mentioned earlier: Felix is homosexual, the cop is a black man and Raj is a non-Western citizen. As minorities, they are closer to the feelings of the sisterhood as oppressed women (both in the Seventies and in the Clone Club as well) because they, as Others, share the feelings of rejection with both, women and clones. Unlike in the Seventies, these Others are welcomed within the Clone Club. Felix is declared a true "sister" in "Ipsa Scientia Potestas." Meanwhile, white Western men such as Beth's boyfriend Paul (Dylan Bruce) are suspected—with good reason—of being undercover agents keeping an eye on the clones. White men are here to prevent and disrupt sisterhood, a situation that resembles the Seventies and the backlash that feminism has had to endure.

Posthumanism and Cloning Are a Girl's Best Friend

Cloning is a politics of posthumanism, a way to enhance human survival and surpass life's boundaries. To many, cloning is a path to the blurring of identity. The way that the body is represented in biomedicine can result in the standardization and commodification of the human body (Lisa Cartwright, p. 40). However, Braidotti feels that our identity was not that concrete and secure to begin with, so the posthumanist offer is a good one.

Braidotti embraces the ambiguous potential that becoming posthuman, more than human, might bring. Braidotti believes that the posthuman can be used as a liberating force that addresses and overcomes "inhuman(e) aspects of our era" (*The Posthuman*, p. 3), with a new ideal of sisterhood based first, in the sharing of likeness and DNA, and second, in inclusive, rather than exclusive, politics. The posthuman can also be a useful tool for understanding women's existence in an age of biotechnological manipulation and genetic alterations.

Thus, although traditional sisterhood has incarnated the notion of the unity that empowers and provides solidarity among its peers, *Orphan Black* seems to play with the idea of mixing or remodeling identity, whereby a "sister" continually exchanges her identity with some other clone, or even giving freedom to the dark side, the "evil twin" within (*Sisters on Screen*, p. 178).

This is related to the fact that the idea of building a unified female voice is no longer feasible; we have to emphasize internal differences. Unlike in the Seventies, the new posthuman feminism does not sweep away the differences, but rather, celebrates them. Motherhood is not an institution of masculine oppression (even when those in charge of the experiments are interested in Sarah because of her fertility, since the other female clones are incapable of giving birth), but a female choice. After all, Sarah's struggle throughout the series is to keep custody of her daughter, to the point that the cliffhanger of Season One is the threat of losing Kira. Motherhood has become a free choice and not a social imposition.

Cloning can be seen as a threat, but here, within the universe of *Orphan Black*, is also a potential return to the politics of sisterhood—now one hundred percent free of exclusions and ill-conceived ideas about homogeneity.

14
Re: Production

DARCI DOLL

The following memoranda have been provided by individuals collaborating with us to expose the activities of the organizations involved with Project Leda and Project Castor. The contents within these documents are significant enough to warrant being shared despite ongoing development of the project. To ensure confidentiality and safety of collaborators, private and identifying information has been redacted. After all, Ipsa Scientia Potestas Est, *yet it's worth limiting the potential transitory sacrifices of this crisis.*

To Hound Nature in Her Wanderings

The data provided by monitors about the clones yields results that are in many ways promising; worrisome in others. In general we find the clones in both Projects Leda and Castor to be extraordinary with respect to skills, intellect, and potential. With only a few exceptions these individuals are productive members of society and offer a lot of promise for the future of the project. Since we have ample evidence of the positive results from the genetic modifications, and their successes warrant little need for further adjustment, the following focuses primarily on the matters of philosophical significance; the areas which we need to fix.

Knowledge of Causes, and Secret Motions of Things

[redacted] explains that the Ancient Greek philosopher Aristotle wrote that something which fulfills its function is capable of achieving excellence; for humans, this means having the potential to achieve *eudaimonia* (commonly translated to mean excellence, happiness, or flourishing). According to Aristotle the function, goal, or best life for humans is the development, expression, and cultivation of rationality; that is, the best life is one of philosophical contemplation.

Aristotle argued that *men* specifically are capable of fulfilling this potential; women, due to their natural reproductive differences are inferior to men and their primary function is reproduction. (There are passages in Aristotle indicating that women may have the ability to attain this ultimate human good, but the more explicit theme is that women are biologically inferior.) The ideal process that Aristotle prescribes is one of contemplating the world around us, learning and understanding what exists, and learning how to use practical wisdom to make decisions as well as to develop good character traits.

Since Aristotle claims that women have been unable to achieve this, the Aristotelian tradition resulted in a devaluation of women. They can't achieve the rational capacities that Aristotle values and are instead evaluated based on their fulfillment of their reproductive function. While philosophers have diverted from Aristotle in many ways, the emphasis on rationality has persisted and with it the assumption that women are inferior to men with respect to rationality. In addition, that women are defined by their reproductive abilities has subsequently (whether or not intentionally) reinforced the alleged inferiority of women.

Formalized, Complex, and Costly

After receiving this explanation of the role of rationality and reproduction in Philosophy [redacted] maintained reg-

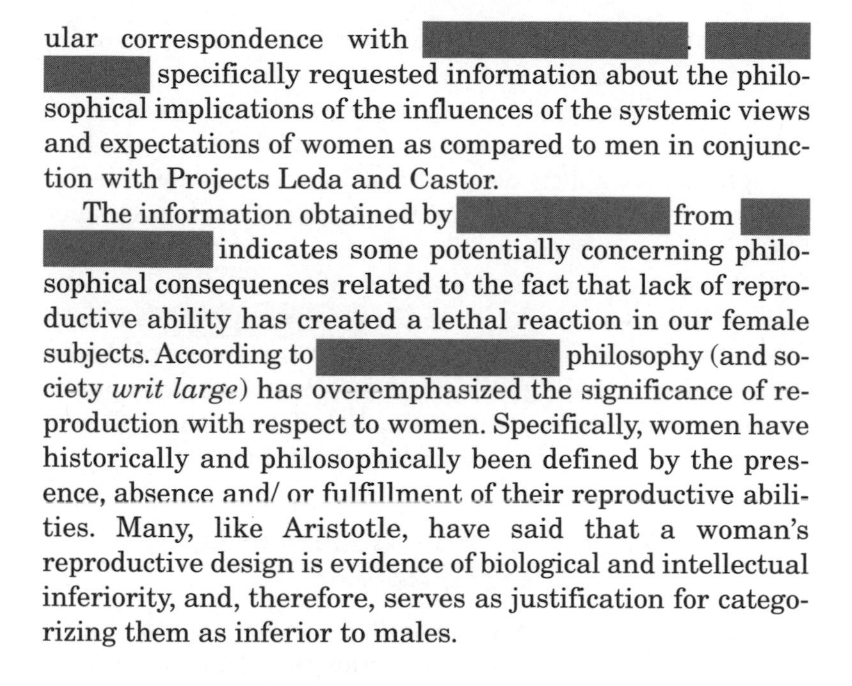

ular correspondence with . specifically requested information about the philosophical implications of the influences of the systemic views and expectations of women as compared to men in conjunction with Projects Leda and Castor.

The information obtained by from indicates some potentially concerning philosophical consequences related to the fact that lack of reproductive ability has created a lethal reaction in our female subjects. According to philosophy (and society *writ large*) has overemphasized the significance of reproduction with respect to women. Specifically, women have historically and philosophically been defined by the presence, absence and/ or fulfillment of their reproductive abilities. Many, like Aristotle, have said that a woman's reproductive design is evidence of biological and intellectual inferiority, and, therefore, serves as justification for categorizing them as inferior to males.

Project Leda: Governed by True Reproduction

The decision to genetically impose infertility in the females has had an undesirable correlation with a fatal illness. While the project scientists are looking into the cause and the development of a cure or vaccination, any potential correlation with the philosophical information provided by is being taken under advisement with regard to any other significant areas of consideration.

Specifically, as we continue the development of clones and research of the lethal defect we want to ensure that we're not also guilty of identifying women as being valued only in conjunction with their reproductive abilities. The female clones Helena and Sarah, and Sarah's daughter Kira may be of particular value for identifying how we can reverse the negative consequences in current and future clones with respect to both the fatal flaw and the way that we identify the role and value of reproductive abilities. This is not to say that their reproductive abilities have to be maintained in order to maintain value; in fact, it's the opposite. We want to ensure that the reproductive autonomy of the clones is respected and restored. However, we should ensure that a lack of reproductive ability does not continue to destroy, or be interpreted as a devaluation of, these exceptional specimens. [As will be discussed in later memos, Helena, Sarah, and Kira have been identified as being of more value than the clones that are infertile] Thus far, ████████'s follow-up research confirms that the general treatment of the female clones, including in some cases their own perceptions of self-worth, demonstrates the belief that the value of women comes from their reproductive abilities.

Project Castor: Governed by Sound Reason

As with our female clones, the male clones are also exhibiting a fatal defect, potentially tied to infertility. However, unlike the females', the males' defect does not develop in the reproductive organs. Instead, it develops in the brain and affects cognitive abilities prior to death. Equally alarming is that female sexual partners of the male clones contract infertility from our males.

████████ notes that this reinforces the philosophical problems identified with Project Leda and demonstrates further philosophical concerns. The philosophical prioritization of the ability to exercise rational abilities and autonomy (making decisions about your own well-being) has often lead to the identification of these as being some of the best features or activities of humans. [As was noted in previous memos]

This same tradition identifies women as being superior with respect to reproduction, but inferior with respect to rationality. Therefore, philosophy has given more value to men than to women. Aristotle, for example, includes a variation of these components as being essential to the ideal human. When we look at the risk posed to the clones in Projects Leda and Castor, it's clear the emphasis is that the former involves a restoration of reproduction, while the latter requires a restoration of rationality. If the rational capacity is threatened or diminished in our clones, we may see the philosophical devaluing of the subjects. An inability to be rational or autonomous may have a correlating assumption of inferiority. Since we focus on the reproductive components of females and intellectual components of the males, we may be guilty of perpetuating this tradition. While concurs with the Philosophers that a threat or compromise to rationality is troublesome, the concerns with Project Castor are much bigger than rationality alone.

The primary focus of the subjects of Project Leda and Castor has been on the fatality of the clones and the success of the project; however, the compromised rational ability in the Project Castor clones has resulted in more significant and dire consequences for society as a whole. These consequences, which do not seem (yet) to have gotten more attention than the other consequences, have resulted in psychosis in the advanced stages that can be a threat to individuals other than the clones. That the concern for the victims of Project Castor is minimal in comparison to the wellbeing of the Leda and Castor clones indicates that Project Castor is being treated preferentially. It should be noted that to date, it's unclear the true cause of the intentional infertility, however, speculates that this could be *at least in part* due to an assumption that in order for clones to maximize their potential they have to be unburdened from reproductive potential. (While this is a potential influence, it's also equally if not more likely that the motivation was to maintain a controlled experience without the variable of the offspring of clones. We don't know yet if the reproductive abilities may be restored when the analysis of the clones is completed. Scientifically, both are significant.)

The Weight of This Combination

The philosophical consequences of the fatal disorders connected with reproduction and rationality in our female and male clones respectively are concerning. As we move forward, we want to ensure that our females aren't defined by reproduction only; especially since so many are superbly productive outside of reproduction. In them, we have a very rare opportunity to challenge the negative stereotyping and dismissal of females. In the remainder of this report the correspondences between [redacted] and [redacted] will provide a philosophical analysis of the treatment of our clones with respect to gender, rationality, and reproduction by reexamining classic philosophers such as Aristotle in conjunction with contemporary feminist philosophers such as Simone de Beauvoir.

Scarred by Many Past Frustrations

[redacted] says that understanding the full effect of this influence may come from examining the work of Simone de Beauvoir, specifically *The Second Sex*. Beauvoir argued that "woman" is often synonymous with "womb;" that women are often identified as being defined by their reproductive abilities. While males are granted a richness of identity beyond reproduction, females aren't afforded that same luxury. Beauvoir argues that we should abandon the reduction of women to biological function, and should extend to them the complex, rich identities men are presumed to be entitled to.

Historically, men, unlike women, have been granted free expression and autonomous self-determination. Being a man isn't fixed or limited by biology. Instead, men are presumed to have higher order rational abilities that grant them the freedom to be defined by their choices instead of by their biology. For Beauvoir, the consequences of this difference in classification of men and women is problematic in its erroneous assumption, as well as in the consequence that she refers to as female enslavement. That is, women are subjected to social oppression, denied legal and social equity, and are diminished with respect to views of social production and contribution. The social expectation for women is that they should reproduce; no further personal or social development is expected.

Unconscious Selection

██████████'s assertion that our clones were subject to these philosophical trends wasn't immediately accepted by ██████████; ██████ understood the history behind these philosophical categorizations based on sex. However, it didn't immediately follow that these correlations were present with respect to the project's treatment of the clones. Upon further reflection, ██████ identified some ways in which the female clones were treated differently than the males which correlated with the problematic trends identified by ████████.

First, the source of fatality in the clones corresponds with the female as reproduction, male as rational dichotomy. The illness in female clones originates in the reproductive organs and spreads to the lungs, eventually killing them. The fatal illness in the male clones, however, stays isolated in the brain. This is reinforcing the view that women without reproduction, and men without rationality, are flawed (and in this case fatally so).

Second, this dichotomy is reinforced by the perceptions of the female clones. The reproductive role of the female clones is prioritized by the clones themselves as well as by the members of Project Leda and Castor. The monitors' reports demonstrate that the female subjects express devastation upon learning of their infertility (this is most predominately documented by the monitors of Alison and Beth). This is in part due to the social conditioning of the prioritizing of reproduction; also in part due to the lack of reproductive autonomy.

█████████ stresses that both males and females should have a right to reproductive autonomy; neither should be defined by reproductive abilities and both should have the right to choose the role reproduction plays for them. The female clones' distress in response to the infertility may in part be caused by this lack of autonomy. However, Projects Leda and Castor treat the infertility differently in ways that go beyond autonomy and reinforce the reproductive gender roles. Mainly, the female clones are treated with passive observation; the male clones are encouraged to copulate with other females to see the extent of their sexually transmitted infertility. Project Castor encourages the males to be active in the acceptance and understanding of their infertility.

The exception to the passive interest in the females' reproductive shortcomings is that Dyad/ Topside, Project Castor, and the Proletheans have little interest in the infertile female clones in comparison to the twins Sarah and Helena; the two clones capable of reproduction. All of these groups have expressed high interest in the female clones capable of reproduction. The Proletheans, for example, divert from their plan to exterminate the female clones and make exceptions for Helena and Sarah. The argument is that these clones are capable of fulfilling their destiny; that is, reproduction. In the same vein, the female clones' treatment of Sarah and Helena seems to be influenced by the ability to reproduce. Helena, an admitted serial killer of clones, is welcomed into the fold once she is identified as Sarah's twin, but especially when she's identified as being capable of reproduction. The unspoken acknowledgment is that there is something special, complete, in Sarah and Helena that the other clones are missing.

Third is the difference in treatment of the clones in Projects Castor and Leda. The clones in Project Castor are self-aware and are actively involved in understanding their role, their illnesses, and the research and science behind their existence. This echoes the philosophical trend of treating men as active, rational beings. The female clones in Leda, however, are kept in the dark of the conditions of their existence and have a passive role in the research. This too echoes the philosophical trend of treating women as passive recipients (analogous to the view that they receive impregnation and are endowed with the duty to give birth). Treating the males as capable of informed rational participation and treating the females as passive, non-informed participants, reinforces the idea that men are expected to fulfill rational and autonomous roles and women are not. This demonstrates the (presumed) unconscious reinforcement of women as the non-autonomous counterpart to males.

Newer Elements of our Defense

[redacted] has suggested that we look to the more favorable components of Aristotle to demonstrate how women are equal to their male counterparts. While Aristotle would technically disagree with the following suggested application of their works, [redacted] feels that the adaptation is more philosophically sound. [redacted] also suggests that we incorporate Simone de Beauvoir to better understand the problematic connection of women being identified as "womb" or "other." Upon completion of [redacted]'s research and investigation, the suggested course of action and analysis of [redacted]'s assessment will be provided.

History Yet to Be Written

[redacted]'s assertion is that our problematic reliance upon the male-autonomy / female-reproduction dichotomy isn't without potential resolution. First and foremost, it's important that we have acknowledged that there is a built-in bias that women are treated differently due to reproductive ability while men are treated differently with respect to rational ability. Having acknowledged that bias, we should focus on the parts that may have been overlooked.

Conditions of Existence

As was previously mentioned, the data on the female clones identifies them to be impressive specimens. They are each capable, competent, intelligent, and versatile. The females who became self-aware were capable of adapting to the environment, and of creating and carrying to fruition a series of complex plans. Their characters seem to be admirable; even Helena was able to overcome her Prolethean training and become a cohesive, productive member of the group.

believes that this is an area in which Aristotle can be re-examined to show the value of the female clones. Aristotle argued that practical wisdom should be used to help us understand context, information, and the world around ourselves. He also argued that this practical wisdom is what enables us to identify, establish, and habituate good character traits, become virtuous, and attain *eudaimonia*. The female clones, then, are evidence that this potential to become a virtuous person isn't determined by biological sex. Moreover, the male clones' disease overrides their rational abilities and makes them a threat to themselves and others. The biological sex of the clones, therefore, doesn't determine the potential for virtue. Instead, it seems to depend upon the combination of one's nature, environment, and exercise of practical wisdom.

Endless Forms Most Beautiful

Finally, asks us to revisit Simone de Beauvoir's call for the free, independent woman. Our investigation has demonstrated that we still view men and women differently with respect to reproduction. It's our task, then to start framing the significance of this and to further the endeavor of equality. We have seen with our clones that their potential isn't limited to reproductive abilities; the limitations imposed upon them with these respects are primarily socially enforced.

What we need to do is find a way to restore equality with respect to consideration of potential and expectations. We need to stop identifying reproduction as the defining criteria and rationality as a secondary property for women. Likewise, we need to stop undermining the importance of reproductive autonomy for men. Both women and men should have the opportunity to fully cultivate their rational abilities, to pursue the lives that they feel are most meaningful, and should be defined by the quality of their character; not by arbitrary assumptions based on sex. By combining the wisdom of Beauvior and the methodology of Aristotle, we may be better equipped to help our clones become the best versions of themselves.

We may be better able to run future interactions or projects. By understanding the full potential of both the female and male clones, independent of reproductive ability, we may be able to actualize that insolvent phantom of tomorrow.

PART V

"You just broke the first rule of Clone Club."

15
The C-Word

RACHEL ROBISON-GREENE

Orphan Black is about more than just cloning. There's a whole lot of technology employed in the *Orphan Black* universe, much of which exists in our own universe as well.

Many of these technologies raise thorny moral questions for which there are no easy answers. For example, contemporary research has made possible (or may soon make possible) each of the following medical procedures: fertilization of an egg outside of the womb, testing for certain kinds of characteristics prior to the birth of a child, and even the selection of certain desired properties in a future child.

Imagine that you're planning to have a child. Presumably you want this child to have the best life that it possibly can. Toward that end, what kinds of things is it morally okay to do for your child before it's born or even conceived?

Suppose that you're choosing a partner with whom to have the child (either in a traditional way or in a new way made possible by technology). Is it acceptable to choose a partner who has all and only traits that you think would most strongly benefit your future child? Would it be permissible to genetically select against certain diseases and disorders while the child is still in the womb? Is it okay to select against certain traits that have traditionally been thought to be disabilities in human beings (for example, blindness and deafness)? Is it acceptable for you, not just to select against certain traits that you think might be problematic for the child, but to actually go a step further to actively select traits that you think would help the child (traits that might make them more successful in terms of intelligence and beauty, for example)?

Knowing what we know about the social stigma that oppressed groups face, would it be permissible to select traits like race, gender, and sexual orientation for the sake of the baby's long term well-being? Would it be acceptable to clone a child to ensure that it has certain traits? Would it be acceptable to genetically engineer a child to have certain traits, and then clone the child so that there will be plenty of people with that set of traits? Would it be okay to do this to create a band of super soldiers that could help us defeat our most threatening enemies?

I presented these questions in the order that might progressively raise the height of your eyebrows. Some of the early suggestions may seem perfectly okay, while you might be very dubious of or even horrified by some of the later ones. How should we carve out this treacherous moral landscape?

The Naturalistic Fallacy

People are often afraid of certain kinds of advances in technology. This may simply be explained by the fact that change is frightening—the ways of living that we have become accustomed to are familiar and comfortable.

Sarah Manning is the same person before and after finding herself to be a clone, but some people might respond better to this new information than others. A further factor may be that we're frightened that certain scientific developments represent a departure from our deeply cherished system of values. Though a departure from our values might constitute a significant threat, it also might not.

If we pause to reflect on the newest developments in technology, we will probably conclude one or more of three things.

First, we may realize that, though our system of values is solid, the new technology does not, in fact, violate that system. For example, *in vitro* fertilization was very controversial when the technology originally emerged, but, once people had a chance to reflect on the matter, most people seemed to realize that the procedure is not actually a violation of their deeply-held values.

Second, we may realize that the advance in question serves as a useful test case for the system of values themselves—it may be that our system of values was either inaccurate or incomplete—it simply took a case of the type we are faced with to illustrate the chink in the value system.

There is, of course, a third option, and that option is that the practice in question truly does violate a well-reasoned system of values that, if given up, would diminish the quality of our lives and the integrity with which they are lived.

Sadly, people tend to be pretty dogmatic in their beliefs; they are often reluctant to subject them—particularly when they concern value judgments—to critical reflection. In other words, people are often unwilling to be philosophical in their thinking, since thinking philosophically is essentially thinking critically about all aspects of our existence, including (and perhaps even *especially*) the aspects that are difficult to challenge. I am quite likely preaching to the choir—the fact that you purchased this book signals that philosophical thinking is important to you (especially when the subject matter is good sci-fi).

Who's Afraid of Big Bad Technology?

Since I take it we have agreed to think critically together, let's go ahead and do some of that. When it comes to technology, what makes people so squeamish?

The first consideration that people frequently raise against developments in technology has to do with the way things are naturally. For example, concerns were raised against the practice of *in vitro* fertilization because it seemed to be a departure from nature. Were the cloning technology from *Orphan Black* fully realized in the real world, we would be likely to see people raising similar objections to the cloning procedure.

When people make claims to the effect that a practice is wrong because it violates the natural way of things, they're committing what is referred to in philosophical circles as the naturalistic fallacy. The fallacy, in the way that I'm referring to it here, is often also referred to as the ought-from-is fallacy. One of the most common examples of the naturalistic fallacy is the familiar argument against homosexual sex. The argument is that because, in nature, sex acts are performed purely for the purposes of reproduction, homosexual sex must be wrong.

This reasoning is bad for a number of reasons, one of the foremost of which is that same-sex sex acts occur in nature all the time. The central premise of the whole argument is a claim that is simply empirically inaccurate. Even if the premise were true, though, we can easily see why these types of arguments go wrong. Their structure is bad—they are invalid, which means that, even if the premises of the argument were true, the conclusion wouldn't follow logically from those premises.

Naturalistic fallacies are frequently leveled, not just against things like homosexuality, but also against new technology that in various ways aids reproduction. Objections to this kind of technology might extend to procedures from *in utero* fertilization, to *in vitro*, to cloning, to eugenics. Understanding the way naturalistic fallacies apply to these sorts of cases will help us to work out some of the tricky questions in *Orphan Black*.

Naturalistic arguments are problematic for at least three reasons. First, there is no reason to think that the way things are is the way that things ought to be. This philosophical principle is commonly expressed using the phrase "You can't get an 'ought' from an 'is'." Claims that use the word "is" are known as *descriptive* claims because they describe the way that the world is. When I say, "The new season of *Orphan Black* is now available," I am simply describing the (albeit immensely improved) state of the world. Before we bring values into the equation, the pure descriptive point about the appearance of the new season tells us nothing about the way that the world should be.

"Ought" statements are what we can call *normative* statements. Normative statements describe not the way the world is, but the way that it ought to be or the way that it should be. Normative statements bring value judgments to the table in a way that descriptive judgments do not. In order to arrive at the claim that we "ought" to be watching the new season of *Orphan Black*, we can't rely descriptive judgments alone. We have to pair descriptive judgments with value judgments about the quality of the show.

There's no reason to think that the way things naturally are is superior to the way that we can make them by using a little bit of precious human ingenuity. Additionally, we should reflect on the fact that (as philosopher David Hume pointed out in a different context), human beings are constantly using their free agency to change the course of nature for their own happiness. Do you find roads useful for getting to and from work? Those didn't simply pop up on their own. Do your eyeglasses help you to see? Those didn't simply grow on a tree. Neither did innovations like flight, the Internet, prosthetic limbs, or any other human-made development. When we appreciate those developments without comment, but criticize other developments on the grounds that they're not "natural," we're engaging in a bit of cherry picking. We're objecting to those developments in the world that make us uncomfortable, but not the technological developments that we personally find useful.

Finally, it's important to recognize that we, ourselves, are *part of nature*. Because we are natural entities, anything that we do must itself be natural. Innovation created by humans, then, is, undeniably *natural*. As a result, it's implausible to think that advances made by natural creatures using natural means to observe and manipulate natural phenomena yield developments that are somehow themselves, unnatural. That point alone doesn't establish that the procedures in question are *ethical*, but the charge that they are unnatural does not stand up to scrutiny.

Once these points are made, it becomes clear that the real concern is not actually that the practices under consideration are unnatural. Instead, the concern is that practices are new and unfamiliar. People are often afraid of things that are new and unfamiliar. Some of these fears are justified, but plenty of them are not. This fear can be the cause of some of the most substantial human evils (think of racism and xenophobia—fear and hatred directed at the foreign and unfamiliar). When we're afraid of the unfamiliar, we need to ask ourselves not just whether the phenomenon we're dealing with is a divergence from the norm, we must also ask ourselves whether it diverges from the norm in a way that is morally bad. At least one way to determine whether or not a practice is morally bad is to look at the consequences that it promotes for creatures that are capable of experiencing pleasure and pain.

If God Had Meant Us To . . .

Plenty of people think that the reason that the world in its natural state is naturally superior is that a superior being—God—created it. For example, Pope Pius XII objected to the practice of *in vitro* fertilization on the grounds those who make use of it, "take the Lord's work into their own hands."

Again, I would appeal to Hume's reasoning. God created humans with the capacity for bringing about stunning innovation. He made possible procedure like *in vitro* fertilization and even cloning. That does not, of course, mean that all scientific developments are ethical, but, if they are unethical, there is some other explanation for *why* that is so. They are not unethical simply because they're departures from the natural state of things.

Can ≠ Should

In *On the Genealogy of Morals,* philosopher Friedrich Nietzsche attempted to find the etymology of the terms "good," "bad," and "evil." He found that, across many languages, the term "good" had ties to words that meant "noble" or something like it, while the term "bad" had ties to words that meant "common," or "low." This suggests to him that the origin of these words came from an early "Master Morality," where "good" traits were natural human excellences.

These would be traits like intelligence, beauty, wealth, power, courage, and the like. According to Master Morality, "bad" traits were simply those traits that were furthest away from the "good" traits. So, traits like stupidity, ugliness, poverty, impotence, cowardice, and the like were understood as "bad."

When I discuss this topic with my students in class, I say to them, "You might find something morally problematic with this list, but who among you, for yourselves, wouldn't trade the later set of traits for the former?" My students, in most semesters unanimously, agree that they would all rather have the traits on the first list than the traits on the second.

If we all universally recognize that the traits on the first list are better than the traits on the second, and we have the option to provide our children with all of the traits on the first list, should we do it? I'm not actually going to take a side on this issue, but I think it's far from obvious that we shouldn't. My husband frequently points out to me that it is never factually inaccurate to answer the question "how did you enjoy "X"? (Where x could be anything: your dinner, your service, your vacation) with the response "I would have preferred it if it were better." Better things are, by definition, better. So, just as you might think God has an obligation to create a better rather than a worse world, we may be obligated, for the sake of our child, to create a better rather than a worse child.

On the other hand, there are serious moral implications for a world-view of this type. The first problem has to do with the potential subjectivity or relativity of the traits that we take to be "good." The set of traits that people in a society value can have serious implications for minority groups within those societies.

First, consider the actions taken by the Nazi party during World War II. Jewish people were considered to be inferior, and on the basis of that perceived inferiority, actions were taken to try to eliminate them from the gene pool, either through forced evacuation or extermination.

There is some reason to be concerned that, given the existence of modern technology, eugenics programs might be actualized (though perhaps less explicitly). Depending on the values of the culture using the technology, the eugenics programs could be either explicit conscious efforts or could be based on implicit, subconscious attitudes. A lengthy study of the history of human kind is not required to arrive at the conclusion that racism has been rampant in pretty much all cultures across time. As a result, if a culture wanted to make a list of desirable traits, that list would probably reflect that culture's attitudes about race (both Leda and Castor are Caucasian and conventionally attractive). If couples or individuals are deciding what traits they want in an engineered child, their choices might be affected by their culture's attitudes about race. This is problematic on both a micro and a macro level. In this case, when I say "micro," I am talking about the decisions made at the level of the individual or couple (or, in the case of *Orphan Black*, at the level of the institution or research team). The potential for creating genetically engineered children allows for the possibility of individual acts of racism. If we all agree that racism is a moral evil, we don't want individuals engaging in acts of racism.

At a macro level, if certain traits are frequently selected for or against, those selections will obviously affect the gene pool. If people select traits that reflect their culture's attitudes about race, there is likely to be less diversity in that culture. You probably already accept that diversity is important, but I'll give you some reasons to think so anyway. Some might argue that diversity is valuable for its own sake. I'm not sure that I buy that argument personally, but it is one of the arguments out there. Another argument is that diversity is instrumentally valuable for everyone. It allows us all to develop our capacities to appreciate new things, learn about the world, and increase our ability to feel empathy in a wider range of cases. Empathy is crucial for morality. Experiences with diversity, then, are of paramount importance for living moral lives—for flourishing as human beings.

Contemporary technology allows us to select for or against traits that are frequently understood as disabilities. Notice that the Leda and Castor clones have full use of all of their senses and mental faculties. To the extent that they have disabilities, those disabilities were either genetically engineered for a purpose, or were outside of the control or understanding of the scientists involved in the project.

The scientists controlling the experiment, at least as far as we have seen so far, have created no clones that are deaf and they have created no clones that have Down syndrome. Some might object that any process that involves selecting against these kinds of traits is guilty of "abilism"—a form of discrimination according to which the traits that people have most commonly are thought of as superior to other, less common, sets of traits. Many deaf people do not consider their deafness to be a disability. Some deaf individuals, when given the opportunity, opt to select for deafness in their own children. They consider themselves to be 'differently-abled' rather than disabled. This is, of course, controversial. Within the limits of this chapter, let's leave it at this—if a scientific advance essentially wipes out an entire group of people in a way that is discriminatory, that scientific advancement is at best seriously morally questionable at worst might be considered downright immoral.

The C-Word

I think that what we have said above, particularly the discussion of the naturalistic fallacy, suggests that there is nothing wrong with cloning as a procedure in general. If there is anything wrong with cloning, it is *the purpose* of the cloning that is potentially immoral, and not the practice itself. In "A Brief History of Cloning" at the end of this book, Richard Greene mentions some common misconceptions about cloning. For example, when you clone a person, the result is not simply an exact copy of the clone with all of the same beliefs, attitudes, tastes in art and music, and so forth. If we cloned Hitler we wouldn't get an army of crazy xenophobes—not unless we trained them that way. So the ethics of cloning is not going to turn on a discussion of whether the person being cloned was evil or not.

Why are Leda and Castor being cloned? Are they being cloned for the pure purpose of bringing life into the world? Are they being cloned for the purpose of providing infertile parents with desperately desired children? Seemingly not. The reasons for their creation are complex, and we get more and more information as the story develops. Neolution's goal is to progress to the next point in human evolution (a goal shared by Nietzsche, whom we discussed earlier). The Ledas and the Castors were created toward that end. And, of course, humans who represent the next stage in human evolution would be ideal candidates for super-soldiers.

We need to look, not at the procedure of cloning itself, but at the negative effects of the procedure as it is implemented in a particular case. If the *only* types of people that are ever cloned are white people like the Leda and Castor clones, the practice, as we have seen above, would be morally wrong. A second moral consideration is that, in the case of the Ledas and the Castors, cloning brings about severe restrictions to privacy and autonomy. They are treated as a mere means to an end, and not as human beings whose autonomy is valuable. This, according to the famous moral philosopher Immanuel Kant, is the very definition of immoral behavior. Their existences are also frequently plagued by serious medical issues which would not be present were they conceived in a more traditional way. (One response to this concern, though, is that Alison, Sarah, Cosima, and Helena would not exist had they not been cloned and they are better off existing than not existing, so the cloning cannot be a bad thing for them. This philosophical problem is known as the non-identity problem and is discussed at length elsewhere in this volume.)

The conclusion that we can draw from this discussion is this—cloning itself isn't bad. It's *how* and *why* you clone that determines the moral status of act. Projects Leda and Castor are immoral—not because they are cloning projects—but because the scientists responsible fail to treat the clones with respect and consideration.

Cloning is fraught with dangers but it is not in itself an evil thing.

16
Is Sarah Manning Responsible for What She Does?

JOSHUA HETER AND JOSEF SIMPSON

Think about Alison Hendrix's drinking problem in *Orphan Black*. A century or more ago, Alison's proclivity to abuse alcohol (perhaps, when triggered by an emotionally overwhelming event) would have been viewed as, largely if not solely, a character flaw, as weakness of will for which Alison is herself responsible.

Addiction science has thankfully progressed far enough for us to realize that there is a deep and significant physiological component to problems like Alison's. Could it be that a century from now, science will have progressed so far that it will be able to tell us the literal and undeniable *cause* of addictive behavior such that an addicted individual was genetically predetermined to abuse?

What is philosophically interesting about this possibility is that genetic, neurophysiological explanations of behavior are not limited to the abuse of controlled substances. Brain science seems to have much to say about a wide range of behaviors, such as aggressive behavior or empathetic response behavior, and about our genetic predisposition to engage in them. The question then becomes, could science progress far enough so that there's no room at all for moral responsibility?

Given the fact that what we do is largely influenced by who we are (genetically, or otherwise), and we do not have control over who we are (genetically or otherwise), can we ever be held morally responsible for what we do? If so, under what conditions?

Genetic Identity and Moral Responsibility

In *Orphan Black*, each of the Project Leda clones, though genetically identical, do not share the same amount of responsibility for their actions. At one end of the spectrum, Helena displays nearly psychopathic tendencies as it does not seem she can recognize why her actions are right or wrong. At the other end, Cosima is consistently deeply conflicted about what's the right thing to do. And so, her recognition that some actions are morally right and others are morally wrong naturally leads us to hold her morally responsible for her actions.

However, since all the Leda clones are genetically identical, the explanation for the (lack of) responsibility does not lie in their genes. This presents a strong argument against genetic determinism in general since, were we to clone human beings, the degree we would hold them responsible for their actions would vary on a case-by-case basis, similar to our intuitive responses to the actions of Leda clones. Some (such as Helena) would be exempt from responsibility and some (such as Cosima) would not be, with a wide range of varying intuitions among those in between. With this in mind, if genetics does not always explain why a person is exempt from responsibility, where else should we look?

Fortunately, when discussing Helena, Cosima points us in a promising direction: "Well, yeah, bad brain chemistry can be genetic, um, but environment, that's individual, right? I mean that's the whole nature-nurture question right there." So, what is it about each of the Leda clones' upbringing that explains our intuitions about their responsibility?

Helena was abused and brainwashed, as was Rachel. Alison was raised by a domineering mother whose standards could never be met and Sarah was a drifter who never settled and always felt angry and felt that the world owed her something. Cosima, by contrast, appears to have been raised in a nurturing and encouraging environment. So, it's presumably the fact that Helena was abused and brainwashed that explains why we think she's not morally responsible for her actions. In a real sense, her actions are not her own. However, attempting to understand responsibility in terms of nurture presents a deep problem.

It's quite natural to think that the reason why the Leda clones act as they do is because of the reasons they have: Helena is conditioned to think the Prolethean view of clones is right and that she is not really killing another person, or when she's killing "actual" people she does so because she feels threatened in some way—thus justifying her actions. Sarah's ultimate reasons for action trace back to her wanting to protect Kira; Cosima is motivated by an intense desire to understand the world and herself, and so on. Of course, each of them has the reasons they do because of their individual beliefs, desires, values, and goals. Let's call these their backgrounds.

So, the Leda clones make the choices they do because of the individual background they each have. This presents a clear picture of how we act: our background makes some reasons for an action more compelling than others and the reasons we adopt (informed by the background) lead to our trying to bring about some action. Various things may come up that stop our action—we may intend to fill a glass of water but fail to do so because the water is temporarily shut off.

As nice as this picture is, it is not without its own difficulties. By appealing to nurture (and the background our environment passes along to us) we are trying to understand why, for example, Cosima is responsible for her actions, but Helena is not. But attempting to understand this in terms of their background beliefs, desires, motivations, and experiences raises the question of why and how their background can transfer (or not transfer) responsibility to them.

If the Leda clones are not responsible for having the background they have, and it's their backgrounds which produce their actions, how could they be responsible for the actions that result? The philosopher Galen Strawson argues that precisely because our backgrounds causes us to act the way we do, moral responsibility is impossible.

Strawson develops his argument like this:

- **We are the way we are, initially, because of heredity and early experience, and these are things for which we can't be held responsible.**
- **We can't at a later stage of life hope to become truly morally responsible for the way we are by trying to change the way we already are.**

This must be true because both the way in which we're motivated to change ourselves and the degree of our success at changing ourselves are determined by how we already are (as a result of heredity and previous experience).

- **Any further changes that we can bring about after our initial changes will in turn be determined by, by way of these first changes, by heredity and previous experience.**
- **This may not be the whole story, because of the influence of random or indeterministic factors. But since we can't be responsible for those random factors, they can't help to make us any more truly responsible for how we are.**

Strawson is arguing here that trying to understand responsibility for action by appealing to our background generates an infinite regress. Put differently, every time we ask, Why did Cosima do *that*?, and we try to answer with something about her motivations or reasons, we can *always* ask, But why does she *believe, think, or value that?* Since each answer also points to something for which she is or is not responsible, we are always led to something for which she is not ultimately responsible. A simpler form of the argument looks like this:

1. **In order to be ultimately responsible for what you do, you must ultimately be responsible for who you are.**

2. **No one is ultimately responsible for who they are.**
3. **Therefore, no one can be ultimately responsible for what they do.**

So, trying to make sense of our intuitions about the responsibility of the Leda clones by appealing to our background experiences, values, beliefs, and motivations seems to lead us to the very odd conclusion that not only are the Leda clones not responsible for their actions, but no one is!

At first, as a way to avoid this extremely counter-intuitive conclusion, we're tempted to think that people *are* actually responsible for who they are. However, just a moment's reflection casts serious doubt on this idea. To be responsible for who we are requires that we actually chose the parents and family we have, the upbringing we happened to have, and all of the various experiences which make up our background.

For example, imagine that Sarah is deathly afraid of dogs and that we can trace this back to her being bitten by a dog as a child. Her fear of dogs affects her behavior by making her avoid them, feeling disgust towards, and so on. The only way she can be responsible for those feelings is if she had chosen to be bitten by the dog when she was a child. If not, all the beliefs and feelings she now has towards dogs, though part of who she is, are not attitudes for which she can be responsible.

Once we realize just how many interactions we have with others and the environment that shaped our background we quickly realize that our actions result from things wholly outside of our control. If we happen to make good moral choices instead of bad ones (like Cosima), we were simply lucky enough to have had the upbringing and experiences that we did. If, on the other hand, like Helena, Rachel, and Sarah, we happen to make really poor moral decisions, we were unlucky enough to have had the upbringing and experiences we have.

So, if there is a problem with this argument it is not in the fact that we are not ultimately responsible for who we are. Perhaps the problem is in the idea of being "ultimately" responsible for our actions and who we are. Why do we need to be ultimately responsible? If we think about ultimate responsibility as meaning total and complete responsibility, then there might be something to this. However, Strawson means something more like "at the end of the day, we can trace back this action you have done to a decision you made without being affected by things outside of your control." However, we have seen that there is no such action. Our actions will always trace back to something—a belief, desire, motivation, value, decision—that we were lucky or unlucky enough to have had.

Genetic Predisposition and Reactive Attitudes

Is this the end of our search for understanding of why some of the Leda clones are responsible for their actions and some are not? Is the answer, contrary to appearances, that none of the Leda clones are responsible for their actions because no one is responsible for their actions? We don't think so.

There are two further ways to explain moral responsibility that appeal to who we are but does not run into these kinds of problems. First, we can turn to Galen Strawson's father for help. Peter Strawson (and many other philosophers today) thinks of responsibility in terms of having the appropriate kinds of attitudes toward others. Second, your upbringing and experiences are important in meeting basic moral standards like the ability to recognize and be responsive to reasons. When your upbringing drastically alters your ability to understand morality, to lack empathy, and so on, then we do not hold that person—such as Helena—responsible for her actions. When your upbringing does not undermine or otherwise drastically alter your ability to recognize and be responsive to moral factors, then we say you have become the appropriate object of moral evaluation—for instance, Alison, Cosima, and to a lesser extent Sarah (though Sarah is probably at the lower end of the standard).

Let us try to understand this a bit more by taking some examples from the Leda clones. we first meet Helena, she is ruthlessly pursuing the rest of the clones. She murders Katja and turns her sights on Beth/Sarah. As viewers, looking in from the outside, we immediately have a visceral reaction towards her. We think she's evil and are disgusted by her actions. More accurately, *she* disgusts us; our disgust is directed not just at her actions, but also at *her*.

However, we quickly find out about her background and the initial attitude is softened, if not changed entirely to pity. As her story unfolds, our attitudes towards Helena change and when Castor captures her, they turn to sadness once we glimpse another depth of her brokenness as she turns to her imaginary helper, Pupok the scorpion.

Here we have an example of the ways our evaluations of another person change once we realize our initial attitudes were inappropriate. Consider an example which doesn't involve one of the clones. When we find out that Donnie is Alison's monitor, it appears justified that we think he is evil. It seems that he, too, is part of the whole conspiracy and has been *maliciously* deceiving Alison for years. But once we learn that he believes he was merely involved in a social science experiment, we find ourselves judging him on his gullibility and stupidity rather than on malicious character. In other words, we again find our attitudes change once we learn more of the character's background.

In both examples, the information we have alters our attitude. This happens in everyday examples common to us all. Suppose someone stomps on your foot and you become justifiably irate. However, were she to inform you that a large black widow spider was on your foot, your attitude would naturally and rightly change to one of gratitude. If this is right, responsibility is more about holding others responsible than about being responsible in some way determined by some absolute standard that one fails (success fully) meets.

Yet, this is not to say that there are no standards at all. Those who maintain that responsibility is a matter of certain kinds of attitudes should also think that those attitudes must be appropriate. It is not appropriate to hold a two-year-old morally responsible for hitting her sister. Nor is it appropriate to hold a thirty-three-year-old who suffers from severe mental disability responsible for her actions. What this tells us is that there are some standards that we think are necessary to meet. So how do we decide who and who is not responsible, which attitudes are appropriate and which not?

The philosopher John Martin Fischer argues that in order to be responsible for our actions we must meet two conditions. First, whatever faculty or mechanism we used to act must be reasons-responsive. In other words, if we decide to do something after deliberating, that ability to deliberate must be so moved by reasons that are different than reasons that would have moved us to do something else. Sarah actually presents a good example here.

After Kira is hit by a car because Helena tried to abduct her, Sarah wants nothing more than to kill Helena. However, upon finding Helena in a cage she is moved to release her. Later, when she is about to hand Helena over to Dr. Leekie of the Dyad institute, she is again moved to instead take Helena home to meet her birth mother. Helena is about to assassinate Rachel, but is convinced not to by Sarah.

This raises an important point about the kinds of reasons involved. It also points to the fact that even those we generally think are exempt from blame and responsibility sometimes do the right thing for the right reasons. The idea is that the reasons that generally move us to action should be those that would motivate most competent people. So the difference between Sarah and Helena is that the reasons that generally motivate Helena to act are reasons that most rational, well-adjusted individuals would not accept. Indeed, most well-adjusted, rational individuals would adamantly reject those reasons.

Again, Sarah is a harder case because she is, in many ways, in moral transformation. But Cosima is a better example since she is mostly motivated by reasons that rational, well-adjusted individuals would accept. The first condition we must meet to be responsible for our actions, then, is to be responsive to reasons that would motivate most well-adjusted, rational people.

The next condition is that the faculties, abilities, or mechanisms we use to act must be ours. They must be a feature of our psychology that we have taken ownership of. To see this we need only consider Helena. Her decisions to act are not hers. She is motivated to act as she does because of the programming she received from Tomas and the Proletheans. Even if some reasons (even reasons that most rational, well-adjusted individuals would accept) would motivate her to do something different, they are taken into consideration by a mental faculty corrupted by abuse and brainwashing.

A central feature of ownership of one's action or the psychological mechanism that caused that action is that the agent sees herself as the appropriate subject of reactive attitudes directed at her because of the way (or the reasons upon which) she acted. In this way, to own a bit of moral psychology is to *take* ownership of it. Clearly, Helena would not see herself as the appropriate subject of disapprobation, in general, let alone for actions she decides to take to further the Prolethean aims.

Contrast Helena with Cosima and the story is quite different. Cosima is throughout deeply concerned for others, and sees herself as rightly being subject to moral evaluation. Similarly with Alison and (to a lesser extent) Sarah. We have arrived at the view that we can be morally responsible for our actions even though we are not ultimately responsible for who we are. And, when we think that someone is not responsible for their actions on the basis of their background beliefs, desires, goals, experiences, and upbringing it is not because our background causes our actions and we are not responsible for our background. Rather, it's because our background can so fundamentally distort our development that we fail to meet the basic standards of moral responsibility: reasons-responsiveness and ownership of our actions (and psychology).

Who's Responsible?

As disciplines such as psychology and neurophysiology progress, we continually discover that there is less and less within our control than we previously thought, and some people conclude that everything that happens is outside our control. Since individuals can't be held morally responsible for what happens outside of their control, this gives rise to the dismal conclusion that no one can be held responsible for anything.

The thought-experiment of *Orphan Black* presents a compelling negative answer. Since we have wildly different intuitions concerning the moral responsibility of the Leda clones, we have reasons to think that our actions are not genetically determined. This led us to attempt to understand why we have the differing intuitions we do. In our examination of this question, we realized that each of the Leda clones had very different backgrounds comprised of experiences, the way they were raised, beliefs, desires, and goals. By reflecting on this it is tempting to conclude that not only are none of the Leda clones responsible for their actions, but no one is.

But to draw that conclusion would be to confuse the idea that our backgrounds cause our actions with the fact that our backgrounds affect our development. Sadly, some individuals are raised in environments or have early experiences that so drastically distort their ability to respond to reasons and take ownership of their actions, that they become exempt from moral responsibility; attitudes of disapprobation, disgust, and blame are simply inappropriate. Of course, if those individuals are dangerous as a result, then we get them off the street, but we simply don't think they are *morally* responsible for their actions.

Thankfully many other individuals do develop to meet basic standards of rationality and moral responsiveness. They can indeed be held responsible for what they do.

17
Dialog with the Buddha

CHRISTOPHER KETCHAM

Translator's Introduction

This is a translation of the *Prolethean Killijijuk* (Prolethean Truth Telling) from the original Labradorean dialect—compiled by the *Inuttut* dictionary—presumably into French by a defrocked Catholic priest, and now by me into English.[1]

The *Prolethean Killijijuk* speaks of a *Hailiggi Nâligak* (holy wise man who calls himself Buddha) who comes out of the north. The *Prolethean Killijijuk* is somewhat vague of the place of origin of the Canadian or 'new' Buddha as he is now known but most believe it is in northern Ontario.[2]

Proletheanā Killijijuk (1. Enlightenment)

I have heard it said that the young man who would be the *Hailiggi Nâligak* after his enlightenment under the tamarack tree by the great lake returned to his village. The Lord spoke thus to an elder of his village. Said the Lord, "The waters of the lake by the tamarack tree were as still as they

---

[1] Inuttut Words from Labrador Virtual Museum: English-Inuttut dictionary

http://www.labradorvirtualmuseum.ca/home/inuttut_dictionary.htm

[2] There were many Buddhas before Gautama. The Canadian north is just as good a place as any for the next.

would have been if they were glass. In the distant I heard the fish as they jumped into the air to catch bugs which had just hatched out in the warm summer's night. I felt no sting of *kittugiak* (mosquitos) but they did sup from me as they do all beasts of the north country."[3]

"The *kittugiak* are our *ukumailutak* (burden or dead weight)," said the elder of his village in agreement.

The Lord continued, "As the *atsanik* (aurora) faded and the dawn emerged, I sat for a moment. The world came to me as it has never before. The shore of the lake, the trees, the sky, and the wind seemed as one. I have become at peace with the world."

"This is a good thing," said the elder, "but you have the look of trouble in your eyes, not the peace you speak of."

The Lord grimaced and continued. "That is so because there are whispers in the air that concern me. They come not from the winds that bring ice, but from the hot breath of south. At first this was like the *tipalak* (stink) of *TuKuk* (death) that is far off—it came and went. Then stronger now it comes to me, not only as I come south, but also as I think more about the meaning of this breath. I have begun to feel the presence others, many others—others who maybe should not be at all. In this hot breath I see births, like *magguliak* (twins), but not *magguliak*—births from mothers or fathers. It feels as if it is a *magguliak najuttik* (twin thing taken), created by some force other than the *tunngak* (roughly karma). No one has died to bring forth these new beings into the world. It is as if the *tunngak*, has been rent by a force of something that is not natural. The *tunngak* calculus of rebirth has been interrupted by this new form of creation that accounts not for the past good and bad *tunngak* of ones who have passed, but rather by the good and bad *tunngak* of ones who still live. This seems so contrary to the flux of nature and rebirth.

I see again and again the words *iliatsuk kuasak*, 'orphan', and 'black'. This black like black ice that hides in plain sight on

[3] Gautama Buddha was also addressed as Lord.

the water, hiding what danger it can be to any beast who steps on the blackness or ventures into its knifelike edges in a bark canoe. In these words, *iliatsuk kuasak*, I see the parentless children hidden from others in plain sight like the black ice."

The elder showed his discomfort at what this *Nâligak* said and commanded him to remain where he stood. The elder went back to the people and spoke with them about his encounter and returned to the *Hailiggi Nâligak*. The elder said, "It disturbs the people to hear what you have just said. They say you are pursued by a *tonngak* (evil spirit). If you stay too long in this place your *tonngak* may come to like this place. You must go." And with that, the *Hailiggi Nâligak* was exiled by his peoples.

Proletheanā Killijijuk (2. The Demon Mara Disguises Himself as a Raven: *Tulugak*)

I have heard it said that when the Lord left his people he came upon demon Mara, the *tonngak* that the elder spoke of, disguised as a *tulugak* (raven) who cackled in a language that the Lord understood.

Said the Lord, "Your temptations have no effect on me *tonngak*. Be gone."

"I won't leave you," said the *tonngak*."

The Lord shrugged and resumed his journey, putting Mara, the *tonngak* disguised as a *tulugak*, out of his mind. The increasingly powerful scent of the *magguliak najuttik* served to guide his journey.

True to his word, the *tulugak* and his minions did not leave the Lord. They followed him, squawking and foraging, and worrying hawks all along the Lord's journey south.

Proletheanā Killijijuk (3. The Doubts of Tomas)

I have heard it said that when the Lord traveled south through the forest trails of Ontario he came to a clearing and climbed over a fence into a pasture connected to a barn and large farmhouse. The *tonngak* and his minions waited silently at the tree line.

A girl-child riding a horse in the field spotted the Lord climbing over the fence, paused, then turned the horse and galloped back to the farmhouse, screaming as she rode. The Lord crossed the field, passing cattle that quietly grazed along the way. As the Lord came to the gate near the barn, two men and two women approached him. The girl-child now hung behind the skirts of the older woman.

"Whoa, stranger," said the older man, "You're trespassing on our land."

The Lord paused at the fence and said, "I have no desire to offend you Tomas, I am on a journey south and do not know where the paths of the forest will take me."

The knot of farm people stepped back abruptly. The older man frowned and lifted a hoe he was carrying in his right hand and said, "How do you know my name? You protest you are a stranger but yet you know me. State your name and your business. Talk fast, mister or I will flog you to within an inch of your life."

The Lord smiled and said, "I am but a Buddha, a *Hailiggi Nâligak* in my native tongue, Tomas, and on a long journey south to learn more about the *magguliak najuttik*. Your name is familiar to me. I mean no harm to anyone here."

"Buddha," laughed the younger man wearing a cowboy hat, "You hear that all; they got a Buddha clone now. Don't that beat all to Hell. Hey, how do they get a Buddha clone from someone who's been dead for, so long?"

The Lord smiled, "Henry, yes I have heard tell of you as well. All Buddhas are born like you and me. I am not a *magguliak najuttik*."

Said Tomas, "What is a 'maggiliac nattickuk' or what you call it. It isn't English or French that I can make out." Tomas gripped his weapon tighter.

The Lord responded, "A clone, like the woman Helena that you have here now."

"Damn it, Tomas" said Henry Johanssen, "I told you they'd come for her."

"State your business and be quick about it," said Tomas.

"Like you," the Lord said, I want to know what clones are. They have changed the *tunngak* forces, what some call karma, in what way that is not yet certain. You call them abomination."

"Of course they're abominations," said Henry.

"Why would they not be abomination and why would you not be a false Buddha, someone who would steal back our clone for your own purposes? Your tongue is sharp, but are your defenses sharp against the likes of us?" said Tomas raising the hoe above his head with its blade pointed directly at the Lord's head.

The Lord then began his *Killijijuk* to the Proletheans, "The *tunngak* forces, what you may know as the karmic forces, underlie everything: all we think, we say, we do. The karmic forces balance good and bad karma when one lives, and determine into what level of existence one will be reborn. No one can erase what one has done, but one can stop doing evil and commit to the good. This is what Gautama Buddha called walking the eightfold path. Through the eightfold path, suffering of being in the world can be undone and enlightenment and an end to ignorance can be obtained. Even one who is truly evil and produces nothing but bad karma can resolve to follow the noble eightfold path and become enlightened. But karma never goes away, good or bad."

"Sounds familiar from what I've heard," said Tomas, "But what does that have to do with clones, and why are you trespassing on our land?"

Said the Lord, "My journey is towards an understanding of what clones mean. You believe that these clones are evil. I am not yet convinced.

Some say that clones are not real persons; that they have no soul. Or worse, their soul is split between two beings for which there can be no good outcome. Neither of these is true. Buddhists do not believe there is a separate self or soul. We change; we grow from small to large and our minds are always thinking new things, so how could there be a separate soul?"

"Blasphemy," said the older woman.

"Yes Bonnie," said the Lord, "For those who believe in Christ. There is no one God in Buddhism. Our *devas* (gods) are limited in their powers and are also mortal. As we believe in no one God, our beliefs differ from the Jews, Christians, or Muslims. Buddhists know only that the karmic forces are always already there and they choose for us what we will be born as in the next life.

What makes the clones different is that they are not of the next life but of the present life. They have not undergone the process of rebirth in order to become. They come directly from another. This is not new to the karmic flux because the bacteria reproduce by splitting themselves into two. There is no father, no mother only the bacterium who comes before and the bacteria who come after. Both look the same. Are the two who were once one the same still? One might have mutated, and even if not, they both live separately and experience the world separately so they are no longer the same. This is a truth your scientists have said, am I not correct?"

"This is going nowhere," said Tomas.

"I want to hear more," said Henry, "Go on and explain more, stranger."

The Lord continued, "I ask, if karma permits the splitting into two of the smallest beings, is it against karma and the forces of nature for people to do the same? Or, are the levels of existence fixed so that some processes do not follow from one level into another? When the bacterium is reborn into a higher level of existence, this ability to split into two, to clone oneself in nature, appears to be lost. However, through some profound manipulation we get human clones."

"Well, you'll get your answers by movin' on," said Tomas.

Bonnie spoke, "That's downright inhospitable of you Tomas. At least we can offer the young man some drink and perhaps some food before he travels on."

"Yes, and I would like to hear more about this idea about rebirth and the process and them bacteria and all that.

You're welcome to break bread with us. What do we call you anyway, . . . Reverend?" said Henry.

Said the Lord, "Call me what you will. I will break bread with you."

"But . . ." grumbled Tomas.

"That's enough, Tomas," said Henry.

Bonnie turned to walk into the house, "We'll break bread with the reverend Buddha here and then he can take his leave," she said back to them. She shooed her older daughter and girl-child before her and spoke clipped commands to them as they ran ahead. Tomas shook his head and walked towards the barn. By the time he had reached the door he had pulled his shirt off to reveal long weals and still bloody scars. Inside the dark barn came the sound of whipping.[4]

Henry touched the rim of his cowboy hat and motioned the Lord through the gate. Together they walked to the house.

The Lord spoke to Henry, "I smell bread and the *maggiliac nattickuk* in the woman Helena you have hidden in the house."

"She's a bit ripe, ain't she" said Henry. "No matter, she's to be taken care of in the ways of the Proletheans soon enough. Come along, Bonnie makes good bread and the milk's fresh."

After they entered the house the *tonngak* Mara disguised as a *tulugak* flew to the porch rail and paced up and down, in front of the picture window.

Proletheanā Killijijuk (4. The Questions of Henry Johanssen)

I have heard it said that the Lord ate bread and drank water with the Proletheans, but politely eschewed any milk or sausage that was offered him.

After the meal, the man, Henry of the Proletheans, directed the *Hailiggi Nâligak* into the drawing room of the

[4] In "Governed by Sound Reason and True Religion," Tomas stands next to Helena, shirt off, whipping his back. He says to Hank, "Old habit."

farmhouse. The women remained in the kitchen to clean the remains of the repast. From outside could be heard the sharp sound of a whip and the guttural cackling of a raven.

Henry spoke, "But it ain't always natural like you think. Now the bull, if he gets too ornery when he mounts the cow, he can maim or even kill her. That's why we help both of them out as we just done in the barn before you came. So how does that fit into what your theories say?"

Said the Lord, "It isn't the process of procreation that is important but that there is an available egg into which the *gandhabba* can flow.[5] The karmic forces capture the flux from the cow who has just passed and then determine into what level of existence it should next be reborn. What we do not know is which cow that has passed will contribute to the calf that will become from your efforts. It could be from a cow that has just died on this farm, or another cow who has died far away, or even something other than a cow. It is difficult to know how the karmic forces will choose."

Said Henry, "What you're saying just don't make sense. How does whatever it is, that gandgabby thing, get from the dying cow to the embryo of the cow I just serviced if it ain't a soul? Something's got to get from here to there, don't it? Mind you I ain't all that familiar with nor do I much get this whole idea of reincarnation or rebirth as you call it, but it's got to have some sort of substance, don't it?"

Said the Lord, "The karmic forces preserve the continuity of identity, but that which passes from the dying one to the embryo is not of the same substance. The dying one's body dissolves but the flux continues because the dying one was not yet enlightened, which means the dying cow will be born again. This dying one's flux craves to become again and the karmic forces help the dying cow's flux find another that is suitable based upon the calculations of the karmic forces. If the dying cow has spread bad karma, it may be reborn into a lower existence. It is difficult to say what the karmic forces will do.

[5] *Gandhabba* is that which flows from the dying one into the embryo where it will be reborn. It isn't a soul but consciousness itself.

There is no separate thing or soul that passes, only identity. This identity is neither substantial nor is it separate from what had been consciousness before in the dying cow and what is becoming consciousness again in your serviced cow's new calf. Your thoughts from moment to moment change, but you do not disappear and reappear. You are a new thinking thing this moment from the one moment before you saw me crossing the field, are you not? Do you feel differently physically? Perhaps not, but your body is also changing over time. With death your body ceases to exist. It is your flux which passes to the next rebirth, carrying all the karma of your past lives with it. Me, I will not be reborn. I have discovered the path to enlightenment."

"Then what will happen to your flux as you call it if you won't be reborn again?" asked Henry.

Said the Lord, "To think that the enlightened one is or is not after dying, to think that the enlightened one both is and is not after dying, to think that the enlightened one neither is nor is not after dying is not right.[6] The enlightened one is free from clinging to being reborn. However, since no enlightened one has reported back after dying, it is not helpful to speculate upon what the state of the enlightened one's flux will be after death. All we can say is that karma *is* and it does not lose or gain anything by the death of an enlightened one."

"Well, I suppose," said Henry, "But let's get off all of that because it's more important to me to understand what it is about these clones that makes them different. So, you go on and tell me more about that."

Said the Lord, "It is grasping, clinging and craving that causes *unsatisfactoriness* in the world. The wanting to be reborn is such a craving. However, the eightfold path I speak of can lead one to enlightenment, which is the end of such cravings.

Take the clone donors, do they not desire, cling to, and crave more of themselves than they can have in this lifetime? Those who give of themselves—their cells—to produce clones

---

[6] Such an argument is called a "tetralemma," a fourfold argument.

are clinging and craving in two ways. First they have the false belief that a self—their self—will also transfer to this new being. This cannot be true. There is no separate self or soul in the host; the same is true for the clone. This thing that begins from the cells that are of another is not outside of the karmic flux, rather it is part of the flux as is any other embryo that becomes a being. The flux is inexhaustible so there is no void created by the clone's birth.

What does happen is that *all* the good and bad karma of the host is copied and added to the embryo's when the embryo is fertilized. This includes all of the host's karma from all of its previous rebirths."[7]

"Okay," said Hank, "But you got yourself a problem. See in your rebirth thing you got going there in your religion, this karmic flux, or whatever you call it, it finds something suitable to be reborn into. So, tell me if I'm right, based on the little I've heard you say on the subject. Say I'm a bad man in this life. Well it's likely that I'm gonna be reborn into something lesser, am I right?"

Said the Lord, "The karmic forces will choose the proper vessel for your rebirth."

"So, the problem we got here is that what gets reborn in the clone is the mother, or father, or whatever it come from. The karmic forces don't get to choose the embryo if the host, as you say, was a saint or a sinner. It goes right back into the thing it comes from but in a different body. So what do you say to that?" said Henry.

Said the Lord, "Only the body is cloned, not the flux. The flux comes from another who has just died, just like the cow you inseminated before we met. The difference is that the clone now has all of the accumulated karma from the host and the one who has died. That is a heavy burden for anyone to bear.

The flux produces a clone without self or soul as was the host. But this clone, though of same body as the host, will be-

[7] Identical Twins are from a single egg which splits in two. The same logic would apply to twins.

come different from the moment it first experiences life. However, should we keep cloning the clone of the clone of the clone . . . the more we clone the clone . . . the more clones will be descendants of the same original host, the more of physical sameness will we have, and the greater the accumulation of karmas from clones who are not reborn in the normal way. Bacteria accumulate little karma; humans much more. Some may think that this cloning is a good thing, especially when we clone the brightest or strongest. They say if we treat them all the same and help them to cling to their strength or their intellect, and train them to do the same—we will build a strong race. I ask, have we not perpetuated *unsatisfactoriness* and the accumulation of karma without possible release?

We then must ask, 'How could someone who has clung to being so hard to have produced a clone ever walk the eightfold path towards enlightenment?' Even if the clone is harvested against the wishes of the host, the host will know it is there. In either case the host has invested too much in the clone to let go. It will be challenging for the clone and its host to stop clinging to each other, or possibly even resenting each other. As you have seen Henry, this clone of yours fights against others who are clones like her. It is uncertain whether the host or the clone can ever become enlightened."

"Sure, sure. Well, I guess I do see the problem you could get from cutting from the cutting from the cutting and so on. They say that the chromosomes of a clone are a bit more raggedy when the person is born than if they was born the regular way. I can imagine that the chromosome of the clone of the clone of the clone is gonna be downright frayed, to where even it wouldn't live so long at all," said Henry.

Said the Lord, "There is more to this. In the time of Gautama Buddha, a woman whose child died ran to the Buddha and begged him to bring her child back to life. The Buddha agreed but told her to go to six homes where death had not occurred and bring from each a mustard seed that they would grind up together into a potion to restore the life of the child. The mother returned crestfallen because she could not find even one house where there had not been a death.

Death is natural and this is paradox of the clone. Would the woman have been happy if Gautama Buddha had said, well woman, we will take a bit of the child's flesh and make you a new child just like the last one. What would she say?"

At that moment Bonnie entered the room, wiping her hands on her apron. "She would say," said Bonnie, "that this would not be the child I lost even if the child looks just like my child. I would not have the same experiences with this child that I had with my dead child, even though they both look the same. I will see the clone and still be sad for my lost child. I might even resent this clone child for trying to imitate my lost child, even though it will be innocent of such thoughts."

The Lord replied, "What the clone does is demonstrate to us all that there is no such thing as separate self or identity. Whether twin or clone, each is different because everyone's experiences are different. The clone may look the same and have the same physical features, but it becomes in its own way even though it carries with it the cumulative karma from its host along with its own. The fact that the woman wants to cling to her dead child is the problem we face by holding on to things that are impermanent and this produces the unsatisfactory need to be born again. Death visits every household. Clinging to this clone brings the need to be reborn. This cycle of rebirth must stop. We can end rebirth through enlightenment. All I offer is the eightfold path for those who wish to follow."

Henry looked to Bonnie and then to the Lord. He said to the Lord, "Through all of this you are saying that while these clones are not things that violate the karmic forces you call them, and are still human, they could become, well, maybe like cattle. We could create a herd of cattle clones. Mind you, each cow's got ideas of its own, but basically they're herd animals and follow each other around. They come and go to the barn just when you want them to once you learn 'em how to do it. It's the calves that got to learn to do these things. So what I'm hearing, if I'm hearing it right at all, that maybe the clones are gonna to be easier to deal with. We can make them do things quicker than we could if we got new ones that are gonna be different all the time. They got born with mostly the same start of a brain, so if we train them the same right from the beginning, we're gonna get the same results.[8] Now that's something, ain't it. And wouldn't it be wonderful if we could train all the human clones the same way? We'd have an army of clones who would do just what we tell 'em to do."

Bonnie said, "But it isn't what we're seeing now. We got Helena and she's as ornery as that bull out there in the back pasture. There's other clones like her that simply aren't like her, praise be to God."

"But that's the whole point," Said Henry, "Whatever they done to train Helena, we could do the same to train her clones. We would make them all like her and we'd have an army that couldn't be defeated. She's smart, crafty, mean, and tough. She's just the kind of muscle we need to bring the Prolethean message to all the people. Well, thank you, Mr. Buddha, or whatever your name is. I expect y'all want to be going now wherever it is y'all are goin'. The road kinda doglegs a kilometer or so to the right once you leave this place and that will get you to a crossroads where you can choose what direction you wanna go. I know you're gonna ask, so I am gonna say it right and clear. You say you can smell Helena our clone. And that may be true, but as I said, she's the kind of clone the Proletheans need right now and I don't want her hearing about no enlightenment or other such nonsense. So you better be on your way."[9]

The Lord did not object. He bowed and thanked Bonnie for the food and took his leave from the Proletheans. As he stepped out into the porch the *tulugak* looked at him and squawked. The *tonngak* disguised as a *tulugak* remained behind. The young Buddha followed the dogleg south and no *tulugak* followed him again.

---

[8] The brain *may* begin from the same cells, but experience will produce different brain structures.

[9] This isn't exactly the direction that Hank goes in, is it?

A Brief History of Cloning

RICHARD GREENE

I first heard about cloning in the seventh grade. I can actually pinpoint the exact moment. It was the last day of school, we were signing one another's yearbooks, and a friend wrote in mine: "To someone who should never be cloned!"

Had I known what it meant, I might have been offended (although, if I am being completely honest, that was not the worst thing someone wrote in my yearbook—kids at that age can be pretty darned mean). Since thinking about my junior high school years tends to cause me to revert back to my thirteen-year-old self, I just wanna say that Steve Reynolds, the person who wrote that I am someone who shouldn't be cloned (and who is still my friend today), is someone who shouldn't be cloned waaaaaay more than I shouldn't be cloned.

At the time I didn't give it much thought (I don't believe that I even asked anyone what the word meant) but a few months later cloning came up again.

My parents took my sister and me to see Woody Allen's movie *Sleeper* (my parents weren't exactly the best judges of what is appropriate viewing material for children—when I was eight years old they had taken me to see *Rosemary's Baby*). In *Sleeper*, which is set two hundred years in the future, the leader of a group of revolutionaries is killed by a bomb, and all that remains of him is his nose. No worries, however, as cloning in the future is a simple procedure. The scientists need to simply take a chunk of the donor's body (a bit or skin, a finger, or in this case, a nose), put it in the cloning machine, and *voilà*, the revolution can continue as planned.

The plot in *Sleeper* relies on two fairly commonly held beliefs about cloning. The first is that the clone will be the same person as the person who was cloned. In *Sleeper* the revolutionaries believe that they will get their leader back. Not just someone who is a whole lot like their leader, but their actual leader. It's as if they can grow the rest of this guy from his nose! The second is that the clone will be like the person who was cloned in every respect. For example, if the person cloned is thirty-seven years old, the resulting clone will "be" thirty-seven years old (his body will be that of a thirty-seven-year-old and he will look like a thirty-seven-year-old). So the revolutionaries won't get a baby version of their leader; they will get a full-fledged adult (presumably produced by some contraption like the one in the opening credits of Monty Python's *The Meaning of Life*, which bangs out fully formed families on a conveyor belt). Moreover, the revolutionaries expect that the clone version of the leader will have all the same memories, beliefs, desires, dispositions, and attitudes as the original leader. Both of these commonly held beliefs about cloning are, of course, false.

While it's not uncommon for science fiction and reality to diverge (it is fiction, after all), we generally prefer that the science in science fiction be scientifically plausible, and when it is not, we like an explanation of why it is not. (Woody Allen gets extra latitude, of course!) One of the things that fans of *Orphan Black* like about the show is that it goes to such great lengths to offer detailed scientific explanations of the processes used in cloning the Leda and Castor lines, and why the two lines have the features they do (neurological disorder in the case of the latter, and inability to reproduce, in the case of the former). This is not to suggest that *Orphan Black* always gets things right. The script writers seem to believe, for instance, that clones would have the same fingerprints. The science tells us otherwise. But interesting parallels between the science of cloning and the fiction of cloning exist. With all this in mind, let's take a look at a timeline of cloning in science and popular culture!

Send in the Clones

The first *artificial* "cloning" (*natural* cloning occurs whenever you get identical twins, triplets, or quads) actually occurred in 1885. The scientist Hans Adolf Eduard Driesch took a two-celled sea urchin embryo and split it into two separate cells. Each cell grew into a full sea urchin.

When most people talk about cloning, they have something different from this in mind. When people think about cloning they are thinking about "animal reproductive cloning"—taking a cell from some animal and producing a whole new, yet nearly identical, animal (as opposed to just separating two embryo cells, as we see in Driesch's experiment). This is possible because of the amazing fact that every cell of an animal contains—in its DNA—a complete blueprint for making that entire animal.

This cloning procedure is also called "Somatic Cell Nuclear Transfer." There is also what is called "gene cloning," and what is called "therapeutic cloning." These involve reproducing segments of genes and stem cells, respectively. Gene cloning and therapeutic cloning are quite interesting and have all sorts of important consequences, but, frankly, are just not as fun to think about (at least not for most laypersons, though it seems clear that gene cloning and therapeutic cloning, down the road, will bear much more greatly on the lives of most people, and, hence, will become very interesting). So what are the early developments in animal reproductive cloning?

Skipping ahead to the early 1950s we get some major breakthroughs involving the transfer of tadpole embryos. Piggybacking on the work of Hans Spemann who in the 1920s showed that Somatic Cell Nuclear Transfer was possible, Robert Briggs and Thomas King transferred the nucleus of a tadpole embryo into a frog egg that had had its nucleus removed. Later in the 1950s John Gurdon was successful in transferring the nucleus of a tadpole intestinal cell into a nucleus-free frog egg. Each of these experiments resulted in a number of tadpole clones. Researchers learned two further things from these experiments: early embryos are more viable for creating successful clones, and cells retain their genetic information, even when being used for different purposes (for example, when an intestinal cell is used as an embryo cell).

In the 1930s we also see the beginning of cloning in science fiction. In 1932 Aldous Huxley published *Brave New World*. A cloning procedure known in the book as "Bokanovsky's Process" details the splitting of embryos leading to genetically identical twins. In *Brave New World* only certain certain classes of persons are cloned, which allows the government to control populations and to further eugenics programs.

In 1939 William F. Temple published the short story "The Four-Sided Triangle" in *Amazing Stories*. The short story eventually became a novel (1949), which, in turn, became a highly influential movie in 1953 (or at least as influential as low-budget 1950s B-movies could manage to be). The movie, *Four Sided Triangle*, which differs only slightly from the novel, tells the story of childhood friends Bill and Robin, who create a machine ("the Reproducer") that can duplicate any object. The lack of scientific detail about just how the Reproducer works is striking. (Think of Descartes's pineal gland here.) Bill and Robin are both in love with their mutual friend Lena. Lena eventually marries Robin, but agrees to let them reproduce her so that there will be a clone to fall in love with Bill.

As was the case in *Sleeper*, Helen (the clone of Lena) was identical in virtually every way to Lena. She was Lena's same age (again, she was not really Lena's age, as she had just been created, but, rather, had the physical attributes of an adult woman), and had all the same psychological features: the same desires and feelings, the same sense of humor. The problem of course is that in virtue of being just like Lena, she, too, loved Robin, and not Bill.

Another early influential novel is A.E. Van Vogt's *The World of Null-A*, which was originally released as a serial in 1945 in *Astounding Science Fiction*. Whereas *Four-Sided Triangle* took place in contemporary times, *The World of Null-A* is a futuristic utopian tale. Here the main character, Gilbert Gosseyn discovers that "he" has died many times, and that whenever it happens a new cloned body is used to host his mind. In *The World of Null-A*, we see all the same features that tend to demarcate early cloning stories: a real lack of scientific detail, the idea that the clone is the same person as the person from whom the clone was cloned (one of the key moments in *The World of Null-A* occurs when Gilbert realizes that he can never die), and the clone has all the same physical properties as its source.

These are just a few of the cloning stories to appear during this era. Between 1930 and 1969 science-fiction magazines, bookstores, and movie-houses were chock-a-block with tales of human-replicators. Other noteworthy titles include Robert Heinlein's novella, "By His Bootstraps," Fletcher Pratt's *Double Jeopardy*, and Theodore L Thomas's and Kate Wilhelms's *The Clone*, worth a mention because it introduces the word "clone" into literature, though, ironically, the story has nothing to do with cloning. The most popular cloning tale of the early era is, of course, Don Siegel's 1956 classic horror movie *Invasion of the Body Snatchers*.

A number of lists of cloning stories also include Philip K. Dick's *Do Androids Dream of Electric Sheep?* and the movie version *Blade Runner*. While this story certainly is influential, and the androids in Dick's story bear some similarity to the pod people of *Invasion of the Body Snatchers*, I've chosen to leave it out of this timeline, as, strictly speaking, it does not involve clones.

In *Invasion of the Body Snatchers* aliens deposit spores in the earth that, once mature, effectively clone the person nearest them by taking on their physical and psychological features (except, of course, their emotions, which these "pod people" lack). They then eliminate the originals in an attempt to take over society. *Invasion of the Body Snatchers* is significant for a number of reasons (lots of social commentary), but for our purposes two things stand out. First, stories are finally starting to provide explanations of how cloning can occur, albeit, the explanations are not all that satisfying ("It's alien pods that are doing the cloning work"). This still has to count as an improvement on "We've invented a machine that can duplicate things." Second, the idea of clones was beginning to become part of the public imagination. Virtually everyone in the United States in the 1950s was terrified of pod people (in the 1950s peoples' imaginations tended to run away with them a bit). Cloning was not just for sci-fi nerds anymore!

Hello, Dolly!

The 1970s brought forth huge advances in animal reproductive cloning, especially the cloning of mammals. Virtually all of the science fiction of the early era, whether it be in print or film, involved the cloning of mammals. In fact, almost all instances involved cloning human beings. This, of course, is the prospect that most non-scientists found simultaneously fascinating and terrifying—fascinating when you get to clone the girl of your dreams, as does Bill in *Four Sided Triangle*, and terrifying when the aliens of *Invasion of the Body Snatchers* do it to replace humans with clones. The 1970s gave us our first steps in this direction.

A big step toward the cloning of mammals was made by the embryologist J. Derek Bromhall in 1975. Bromhall transferred the nucleus from a rabbit embryo cell into a rabbit egg cell from which the nucleus had been removed.

In 1978 there appeared the sensational and hugely popular book *In His Image: The Cloning of a Man* by David Rorvik, a then reputable medical and science journalist. In the book and in media interviews, Rorvik claimed that he had been contacted by a reclusive millionaire known as 'Max' to put together a scientific team to make a clone of Max. The project had been a success, and Max's clone (a young child, of course) was alive.

Most scientists were skeptical since they believed that cloning techniques were insufficiently advanced and that the cloning of any mammal was far in the future. Also, there were major difficulties in the way of developing an embryo from a cell of an older mammal—and Max was supposedly in his sixties.

Then J. Derek Bromhall brought a seven million dollar defamation suit against Rorvik's publisher. Bromhall maintained that *In His Image* drew upon his work without his permission and was a hoax which damaged his reputation. The publisher paid Bromhall $100,000 and declared the book a hoax, but Rorvik maintains to this day that it is all true. Most people now think that *In His Image* is science fiction masquerading as science fact, but there are still some believers.

In the 1980s experiments yielded the birth of cloned mammals. Bromhall's experiment had ended without attempting to bring the fertilized rabbit embryo to birth. In 1985 Steen Willadsen was able to produce three live lambs by fusing a cell from a sheep embryo with an egg cell containing no nucleus. In 1987 Neal First, Randal Prather, and Willard Eyestone were able to do the same thing with cows. While these were huge steps forward, they don't quite count as cloning in the sense that most people understand the term. That's because these experiments begin with embryo cells, and the goal is ultimately to produce clones from other cells—any other cells.

The 1990s yielded the true cloning of mammals. In 1996 the Scottish researchers Ian Wilmut and Keith Campbell place a cell's nucleus (taken from an adult sheep) into a sheep egg cell (again, one without a nucleus). The result was Dolly—the very first cloned mammal. Dolly became a worldwide phenomenon. Suddenly everyone was aware of cloning.

In 1997 Li Meng, John Ely, Richard Stouffer, and Don Wolf, using methods similar to Steen Willadsen, were able to clone two primates. They were rhesus monkeys named Ditto and Neti. This gets us one step closer to cloning humans.

Later in 1997 the team that brought us Dolly, along with scientist Angelike Schnieke, brought us Polly. A second sheep. Polly, however, came from cells that had been genetically engineered so that her milk had therapeutic properties (in this case it was engineered to reduce blood clots).

The next big step towards the cloning of humans occurred in 2007. Scientist Shoukhrat Mitalipov's research team was able to create advanced primate (again, a rhesus monkey) embryos using adult cells. The resulting cells were embryonic stem cells. Embryonic stem cells have the unique feature that they can become whatever type of cell they are needed to be, and hence, are useful for treating diseases. In 2013 the same research team was able to produce human embryonic stem cells.

Despite a number of claims to the contrary by various scientists in South Korea (ultimately such claims have been disproven or retracted), there has yet to be a successful and generally attested cloning of a human being. But it's only a matter of time. For reasons of selfish vanity, we hope it doesn't happen while this book is being printed.

It's not surprising that cloning in popular culture also made huge advances during this later era. In addition to devoting more time to making the details of the various cloning processes explicit, the later era cloning stories raise a variety of ethical and metaphysical issues, which the earlier stories either neglected or gave only a cursory treatment.

Our Brave New World of Cloning

Earlier cloning stories pretty much focused on what can go wrong when cloning occurs (clones take over the world, or they fall in love with the wrong person), whereas later cloning stories tackle tough questions about a whole host of issues ranging from eugenics, social order, poverty, personal identity, and humankind's relation to nature (just to name a few). In recent years (since news of Dolly the sheep went viral) there has been great interest in cloning, which, in turn, has led to even greater proliferation of novels, short stories, plays, and movies about cloning. Here are just a handful of the more influential ones.

The contemporary era of cloning stories gets ushered in by Ira Levin's 1976 novel *The Boys from Brazil*, which was adapted for film in 1978. Here we move beyond the simplistic "clones are attacking!" type story line, and replace it with a compelling study of what would happen if someone really evil (in this case Adolf Hitler) were cloned. While *The Boys of Brazil* is somewhat lacking in scientific detail, it does nicely present a complex relationship between genetics and environmental factors in determining behavior (in the story, in order to guarantee that the Hitler clones will end up with Hitler's particular personality traits, his creator, Josef Mengele, goes to great lengths to ensure that each experiences similar major life events to Hitler, such as the death of his father at the age of thirteen).

The first cloning story to present a highly detailed scientific account of how cloning can occur is the 1990 novel by Michael Crichton *Jurassic Park*, which was made into a vastly successful film in 1993. In Jurassic Park, extinct species of dinosaurs are brought back via somatic cell nuclear transfer. The genetic material was taken from the dinosaur blood found in Jurassic-era mosquitos that had been preserved after being trapped in amber from trees. The genetic codes from the dinosaurs was largely intact, and gaps in their genetic sequences were provided by frogs.

Due to the popularity of *Jurassic Park*, the scientific details of cloning became demystified for fans of the book and movie; suddenly everyone knew how cloning worked. Subsequent stories no longer needed to provide the details of the experiments to their audience, unless doing so explained some nuanced feature essential to that particular story. *Jurassic Park* was perhaps the first cloning story to really grapple with questions surrounding the permissibility of cloning. Prior to its publication, cloning was pretty much assumed to be a bad thing.

The 1996 film *Multiplicity* takes a light-hearted look at cloning. It's not the first film to do this, as *Sleeper* did the same thing twenty-three years earlier, but here cloning is at the forefront the entire time.

Influential films that continue the tradition of exploring the social and moral implications of cloning are 1997's *The Fifth Element* (pretty much only mentioned here because of the $260 million dollars it raised at the box office), 2000's *The 6th Day*, 2002's *Star Wars: Episode II—Attack of the Clones*, and 2012's *Cloud Atlas* (based on the 2002 novel). Of particular note is 2005's *The Island*, which raises related issues regarding clone value and clone rights (in *The Island* clones are created to provide organs for wealthy people.) Similar themes are considered in Kazuo Ishiguro's 2005 novel *Never Let Me Go*, which was adapted for film in 2010.

2009 brought two more influential films that placed cloning at the forefront: *Moon* and *Splice*. In the former the clone is the protagonist (this is not the first time this occurred, but it is particularly effective in *Moon*). In the latter, cloning occurs to bring about human-animal hybrids. At this point we've come quite a long way since the cloning in Huxley's *Brave New World*.

This brings us to 2013 when *Orphan Black* appeared on the scene. One of the things that fans like about *Orphan Black* is that the details of the cloning experiments are central to the plots and characters in the show. It's not just a show about how cloning works, and it's not just a show that assumes that cloning works. It's a show about the different ways in which cloning can work, and more importantly, it's a show about the different ways in which cloning can go wrong. There is much more to say on this topic, but for that you'll have to keep on watching *Orphan Black* and read the rest of this book!

References

Ang-Lygate, Magdalene, Chris Corrin, and Millsom S. Henry. 1997. *Desperately Seeking Sisterhood: Still Challenging and Building*. Taylor and Francis.

Aristotle. 2016. *Metaphysics*. Hackett.

Bammer, Angelika. 1991. *Partial Visions: Feminism and Utopianism in the 1970s*. Routledge.

Beauvoir, Simone de. 1976. *The Ethics of Ambiguity*. Citadel.

Benatar, David. 2006. *Better Never to Have Been: The Harm of Coming into Existence*. Oxford University Press.

Boonin, David. 2014. *The Non-Identity Problem and the Ethics of Future People*. Oxford University Press.

Braidotti, Rosi. 2011. *Nomadic Subjects: Embodiment and Sexual Difference in Contemporary Feminist Theory*. Columbia University Press.

———. 2013. *The Posthuman*. Polity.

Camporesi, Silvia. 2009. Choosing Deafness with Preimplantation Genetic Diagnosis: An Ethical Way to Carry on a Cultural Bloodline? <www.researchgate.net/publication/40756335_Choosing_Deafness_with_Preimplantation_Genetic_Diagnosis_An_Ethical_Way_to_Carry_on_a_Cultural_Bloodline>.

Camus, Albert. 1991. *The Myth of Sisyphus and Other Essays*. Vintage.

Carter, Robert E. 2013. *The Kyoto School: An introduction*. State University of New York Press.

Cartwright, Lisa 1998. A Cultural Anatomy of the Visible Human Project. In Treichler, Cartwright, and Penley 1998.

Conboy, Katie, Nadia Medina, and Sarah Stanbury. 1997. Introduction. In Katie Conboy, Nadia Medina, and Sarah Stanbury, eds., *Writing on the Body: Female Embodiment and Feminist Theory*. Columbia University Press.

Descartes, René. 2009. *Principles of Philosophy*. SMK Books.

Durkheim, Emile. 1982. *The Rules of Sociological Method and Selected Texts on Sociology and Its Method*. The Free Press.

Epicurus. 2012. *The Art of Happiness*. Penguin.

Faludi, Susan. 2006. *Backlash: The Undeclared War Against American Women*. Random House.

Fischer, J.M. 1999. Recent Work on Moral Responsibility. *Ethics* 110 (October).

———. 2006. *My Way: Essays on Moral Responsibility*. Oxford University Press.

———. 2011. *Deep Control: Essays on Free Will and Value*. Oxford University Press.

Freeman, Jo, and Victoria Johnson. 1999. *Waves of Protest: Social Movements Since the Sixties*. Rowman and Littlefield.

Gamble, Sara. 2001. Postfeminism. In Sara Gamble, ed. 2001. *The Routledge Companion to Feminism and Postfeminism*. Routledge.

Genetic Science Learning Center. The History of Cloning <http://learn.genetics.utah.edu/content/cloning/clonezone>.

Great Britain, Human Fertilisation and Embryology Act of 2008 <www.legislation.gov.uk/ukpga/2008/22/section/14 , accessed 13 Dec. 2015>.

Grebowicz, Margaret, ed. 2007. *SciFi in the Mind's Eye: Reading Science through Science Fiction*. Open Court.

Hollows, Joanne. 2000. *Feminism, Femininity, and Popular Culture*. Manchester University Press.

Hume, David. 1985. *A Treatise of Human Nature*. Penguin.

———. 1999. *Writings on Religion*. Open Court.

Irigaray, Luce. 1985. *Speculum of the Other Woman*. Cornell University Press.

Kamitsuka, Margaret. 2006. *Feminist Theology and the Challenge of Difference*. Oxford University Press.

Klein, Renate, and Susan Hawthorne. 1997. Reclaiming Sisterhood: Radical Feminism as an Antidote to Theoretical and Embodied Fragmentation of Women. In Ang-Lygate, Corrin, and Henry 1997.

Lacan, Jacques. 1998. *The Seminar of Jacques Lacan: The Four Fundamental Concepts of Psychoanalysis*. Norton.

Levi-Strauss, Claude. 1995. *Myth and Meaning: Cracking the Code of Culture*. Schocken.

Locke, John. 1998. *An Essay Concerning Human Understanding*. Penguin.

McRobbie, Angela. 2009. *The Aftermath of Feminism: Gender, Culture and Social Change*. Sage.

Mill, John Stuart. 1988. *The Logic of the Moral Sciences*. Open Court.

Mukherjee, Siddhartha. 2016. *The Gene: An Intimate History*. Scribner's.

National Human Genome Research Institute. 2016. Cloning Fact Sheet <www.genome.gov/25020028/cloning-fact-sheet>.

Nishida Kitaro. 1993. *Last Writings: Nothingness and the Religious World View*. University of Hawaii Press.

Palmer, Brian. 2012. Can Identical Twins Get Away with Murder? *Slate*. <www.slate.com/articles/news_and_politics/explainer/2012/08/true_crime_with_twins_can_identical_twins_get_away_with_murder_.html>.

Parfit, Derek. 1984. *Reasons and Persons*. Oxford University Press.

Prebish, Charles S., and Damien Keown. 2006. *Introducing Buddhism*. Routledge.

Rantala, M.L., and Arthur J. Milgram, eds. 1999, *Cloning: For and Against*. Open Court.

Rorvik, David. 1978. *In His Image: The Cloning of a Man*. Lippincott.

Rueschmann, Eva. 2000. *Sisters on Screen: Siblings in Contemporary Cinema*. Temple University Press.

Sartre, Jean-Paul. 2007. *Existentialism Is a Humanism*. Yale University Press.

Saussure, Ferdinand de. 1998. *Course in General Linguistics*. Open Court.

Seiichi, Yagi, and Leonard Swidler. 1990. *A Bridge to Buddhist-Christian Dialogue*. Paulist Press.

SFE: The Encyclopedia of Science Fiction. 2016. Clones <www.sf-encyclopedia.com/entry/clones>.

Spinoza, Benedict de. 2005. *Ethics*. Penguin.

———. 2007. *Principles of Cartesian Philosophy*. Philosophical Library.

Sturtevant, A.H. 2001. *A History of Genetics*. Cold Spring Harbor Laboratory Press.

Strawson, Galen. 1994. The Basic Argument. *Philosophical Studies* 75. Reprinted in Watson 2003.

Taylor, Verta, and Nancy Whittier. 1999. Collective Identity in Social Movements Communities: Lesbian Feminism Mobilization. In Freeman and Johnson 1999.

Treichler, Paula A., Lisa Cartwright, and Constance Penley, eds. 1998. *The Visible Woman. Imaging Technologies, Gender, and Science*. New York University Press.

Turner, Stephanie S. 2007. Clone Mothers and Others: Uncanny Families. In Grebowicz 2007.

Wargo, Robert J.J. 2005. *The Logic of Nothingness: A Study of Nishida Kitaro*. University of Hawaii Press.

Watson, Gary, ed. 2003. *Free Will*. Oxford University Press.

Weber, Max. 1946. *From Max Weber: Essays in Sociology*. Oxford University Press.

Weinberg, Rivka. 2015. *The Risk of a Lifetime: How, When, and Why Procreation May Be Permissible*. Oxford University Press.

Wikipedia. Cloning <https://en.wikipedia.org/wiki/Cloning>.

The Clone Club

Emiliano Aguilar is a Bachelor in Arts graduated from the Universidad de Buenos Aires (UBA)—Facultad de Filosofía y Letras (Argentina). He is a member of the research group in horror movies "Grite" and has published about science-fiction in journals such as *Lindes* and *Letraceluloide*. He is looking to clone himself six times so that he only has to work one day a week.

Erik Baldwin is Adjunct Professor in the Philosophy Department at the University of Notre Dame. He has written chapters for *Battlestar Galactica and Philosophy*, *Heroes and Philosophy*, and *The Ultimate Game of Thrones and Philosophy*. His work focuses on philosophy of religion, epistemology, and comparative and cross-cultural philosophy, with an emphasis on Christian, Buddhist, and Islamic philosophy. Although he's not as good as the Leda clones, he's been known to rock a few dance parties in his day.

Adam Barkman is an associate professor of philosophy and chair of the philosophy department at Redeemer University College, Ontario. He is the author or co-editor of nine books, most recently *Making Sense of Islamic Art and Architecture* (2015) and *Downton Abbey and Philosophy* (2015). While Barkman is convinced that the world has quite enough Adam Barkmans with just the one, he does think that the world would be improved with a number of clones of his amazing wife, Ashley.

Fernando Gabriel Pagnoni Berns currently works at Universidad de Buenos Aires (UBA) - Facultad de Filosofía y Letras (Argentina), as a graduate teaching assistant. He teaches seminars on American Horror Cinema and Euro Horror. He has published chapters in the books *Horrors of War: The Undead on the Battlefield*, edited by Cynthia Miller, *To See the Saw Movies: Essays on Torture Porn* and *Post 9/11 Horror*, edited by John Wallis, *For His Eyes Only: The Women of James Bond*, edited by Lisa Funnell, *Dreamscapes in Italian Cinema*, edited by Francesco Pascuzzi, *Reading Richard Matheson: A Critical Survey*, edited by Cheyenne Mathews, *Time-Travel Television*, edited by Sherry Ginn, *James Bond and Popular Culture*, edited by Michele Brittany, among others. He wants to live long enough to see the clone of the mammoth that scientists promised long time ago. No news yet. The clock's ticking.

Rod Carveth is Director of Graduate Studies at Morgan State University in Baltimore, Maryland. He writes a lot on the intersection of popular culture and ethics. He is the co-editor of two volumes—*Mad Men and Philosophy* and *Justified and Philosophy*, and has contributed chapters to *Boardwalk Empire and Philosophy*, *The Good Wife and Philosophy*, and *Orange Is the New Black and Philosophy*. Rod is not particularly funny, but his clone is.

Audrey Delamont is a clone monitor, who also happens to be married to her subject. Her husband, the clone, is still under the impression that he is an 'identical twin' and is still completely unaware of the various tests that are regularly preformed on him. Audrey's cover includes having a BA and MA in philosophy from the University of Calgary, working specifically in epistemology and on defenses of skepticism. She has been in deep cover for the past ten years or so, and has grown to enjoy the various aspects of her facade as a monitor, including: her dogs, moonlighting as a watercolor and digital artist, and never being able to pass up the opportunity for a good pun.

Darci Doll spent the first part of her life in the black, unaware of the systemic influences that were dictating people's lives and potential. Once exposed to philosophy she became "self-aware" so to speak, and she began dedicating her life to helping her sestras and sestra-brothers pursue philosophical enlightenment. Darci teaches philosophy at Delta College in Michigan, where she teaches her students to do totally crazy philosophy so that they can win at philosophy.

SARAH K. DONOVAN is an associate professor in the Department of Philosophy and Religious Studies at Wagner College. Her teaching and research interests include feminist, social, moral, and Continental philosophy. As a mother herself, she humbly salutes the tough moms of *Orphan Black* like Sarah, Alison, and Mrs. S. She *might* consider including Helena, although she would be terrified to find herself chatting with her on the playground about parenting techniques.

CHARLENE ELSBY is an Assistant Professor at Indiana University-Purdue University, Fort Wayne. There she monitors the results of her experimental injections of existential dread to an unwitting student population. She is the co-editor of *Essays on Aesthetic Genesis* (2016) and has contributed to several philosophy and popular culture volumes.

RICHARD GREENE is Professor of Philosophy at Weber State University. He also serves as Executive Chair of the Intercollegiate Ethics Bowl. He's co-edited a number of books on pop culture and philosophy including *The Princess Bride and Philosophy*, *Dexter and Philosophy, Quentin Tarantino and Philosophy*, *Boardwalk Empire and Philosophy*, and *The Sopranos and Philosophy*. Richard has reason to believe that he was cloned from the Pillsbury Dough Boy.

JOSHUA HETER, who is definitely not a clone of Josef Thomas Simpson, earned his MA in philosophy from Western Michigan University, and his PhD in philosophy from Saint Louis University. Currently, he is an instructor of philosophy at Iowa Western Community College where he spends much of his time considering the ways his life is (or is not) predetermined by his genetic identity.

JEREMY HEUSLEIN is a doctoral candidate at KU Leuven in Belgium. His research work is on the phenomenology of the body, specifically in the experiences of weightiness and the subject's connection to the Earth. When not reading and writing, he enjoys procrastinating by immersing himself in the worlds of science fiction. Being a middle child, he's always imagined there was a twin of himself out there with whom one day he could reconnect; he hopes there's only one and that he is not trying to kill him.

JOHN V. KARAVITIS. In the time-honored American tradition of F***/Marry/Kill, John V. Karavitis, CPA, MBA, would like to confirm that he would definitely F*** (corporate uber-b*tch) Rachel; Marry (uptight soccer mom) Allison; and Kill (batsh*t insane) Helena. Curiously, John has no feelings whatsoever for *Orphan Black*'s protagonist, Sarah. ("Holy freaking Christmas cake!" Please don't show this bio to Helena. Or to Sarah. Please!)

CHRISTOPHER KETCHAM earned his doctorate at the University of Texas at Austin. He teaches business and ethics for the University of Houston Downtown. His research interests are risk management, applied ethics, social justice, and East-West comparative philosophy. He hopes that the Proletheans do not take to heart the old Asian koan, "If you meet the Buddha on the road, . . . Kill him!" Otherwise, what is left to be done?

ROB LUZECKY is a lecturer at Indiana University-Purdue University Fort Wayne. He has co-written chapters in various Popular Culture and Philosophy volumes and is currently co-editing a book on John Waters and philosophy. When he is not busy catching up on TV and avoiding the nefarious agents of Dyad, he spends his time talking about ethics and the ever-so-cool insights of realist phenomenologists.

DANIEL P. MALLOY has been teaching philosophy and writing about philosophy and popular culture for a while now. He's published chapters on *Star Wars*, *Inception*, *The Terminator*, Batman, Superman, Green Lantern, Iron Man, Spider-Man, and The Avengers. Clones don't worry him. Clowns, on the other hand . . . or cloned clowns . . . or clowned clones . . . (shudder).

RACHEL ROBISON-GREENE is a PhD Candidate in Philosophy at UMass Amherst. She is co-editor of *The Golden Compass and Philosophy*, *Dexter and Philosophy*, *Boardwalk Empire and Phi-*

losophy, *Girls and Philosophy, Orange Is the New Black and Philosophy*, and *The Princess Bride and Philosophy*. She has contributed chapters to *Quentin Tarantino and Philosophy*, *The Legend of Zelda and Philosophy*, *Zombies, Vampires, and Philosophy*, and *The Walking Dead and Philosophy*. Rachel's motto is be Helena on the streets and Alison between the sheets.

JOSEF SIMPSON, who is definitely not a clone of Joshua S. Heter, has been thinking about responsibility for action and belief for over ten years. His curiosity has taken him from southern California to New York, where he fortunately never ran into conspiratorial French graduate students while completing his MA in philosophy at Fordham University. After that, he moved to Washington, DC, where he is currently finishing his dissertation on responsibility and competence at Johns Hopkins University. So far research in philosophy has not attracted attention from Neolutionists, off-book army projects, or right-wing fanaticism so his work will not likely lead to insidious global influence, fundamentalist religious zealotry, military cover-ups, or assassination, but it *might* be more interesting than the small talk heard at suburban parties.

JOHANNA WOLFERT is an independent scholar and dedicated sestra in spirit. Her passion for philosophy rivals Helena's love of food.

CARMEN WRIGHT recently completed her Master of Arts at the University of Rochester where she somehow convinced the administration to let her write her thesis on *Orphan Black*, personhood, gender performance, and queer families where her chapter originated. Ever since the series introduced Helena, she has finds herself slipping into a Ukrainian accent and referring to her sisters as "sestras."

Index

AND PHILOSOPHY

BIGGER ON THE INSIDE

EDITED BY COURTLAND LEWIS AND PAULA SMITHKA

ALSO FROM OPEN COURT

Doctor Who and Philosophy

Bigger on the Inside

VOLUME 55 IN THE OPEN COURT SERIES,
POPULAR CULTURE AND PHILOSOPHY®

In *Doctor Who and Philosophy*, a crew of mostly human travelers explore the deeper issues raised by the Doctor's mind-blowing adventures. They look at the ethics of a universe with millions of intelligent species, what makes one life-form more important than another, whether time travelers can change history, and how the *Doctor Who* TV show is changing the world we live in. The chapters draw freely on both the classic series (1963–1989) and the current series (2005–).

"A fun, informative volume for folks interested in an introduction to philosophy through the vortex of Doctor Who.*"*

— LYNNE M. THOMAS, co-editor of *Chicks Dig Time Lords*

"Lewis and Smithka have done all sapient species a brilliant service by introducing Doctor Who and Philosophy *into the time continuum. Like the Doctor's human companions, we get to travel through a universe of Big Ideas with a caring, clever, and, yes, conflicted friend."*

— PATRICK D. HOPKINS, editor of *Sex/Machine*

"No series in the entire history of television has lit up all the beacons of classic philosophy like Doctor Who, *and this brilliant book is chock full of Time Lord enlightenment."*

— ROBERT ARP, co-editor of *Breaking Bad and Philosophy*

AVAILABLE FROM BOOKSTORES OR ONLINE BOOKSELLERS

For more information on Open Court books,
in printed or e-book format, go to
www.opencourtbooks.com.